Microcirculation - Updates in the Next Frontier of Vascular Disease

Edited by Aleksandar Kibel and Michael S. Firstenberg

Published in London, United Kingdom

Microcirculation - Updates in the Next Frontier of Vascular Disease
http://dx.doi.org/10.5772/intechopen.1001544
Edited by Aleksandar Kibel and Michael S. Firstenberg

Contributors
Marie Jirkovská, Gordon Ogweno, Edwin Murungi, Yuki Matsumoto, Gayathri Victoria Balasubramanian, Roozbeh Naemi, Sharath Kommu, Shalini Arepally

First published in London, United Kingdom, 2024 by IntechOpen
IntechOpen is the global imprint of INTECHOPEN LIMITED, registered in England and Wales, registration number: 11086078, 5 Princes Gate Court, London, SW7 2QJ, United Kingdom

British Library Cataloguing-in-Publication Data
A catalogue record for this book is available from the British Library

Additional hard and PDF copies can be obtained from orders@intechopen.com

Microcirculation - Updates in the Next Frontier of Vascular Disease
Edited by Aleksandar Kibel and Michael S. Firstenberg
p. cm.
Print ISBN 978-1-83769-511-9
Online ISBN 978-1-83769-510-2
eBook (PDF) ISBN 978-1-83769-512-6

Meet the editors

Aleksandar Kibel is an Associate Professor of Physiology in the Department of Physiology and Immunology, Faculty of Medicine, University of Osijek, Croatia, as well as an Associate Professor of Internal Medicine in the Department of Clinical Medicine, Faculty of Dental Medicine and Health, University of Osijek. He is also an internal medicine specialist and cardiology subspecialist at the Department for Heart and Vascular Diseases, Osijek University Hospital Centre, Croatia. He obtained an MD from the Faculty of Medicine, University of Osijek, in 2009, where his research activities included laboratory practicums and research visits at several foreign institutions, including the University Clinic Hamburg-Eppendorf, Germany; the Medical College of Wisconsin, USA; and Chulalongkorn University, Thailand. He obtained his Ph.D. from the Department of Physiology and Immunology with a focus on vascular physiology and pathophysiology. He teaches at the high school, university, and postgraduate levels. He was a mentor for several diploma works. His prior work experience includes work at the Emergency Medicine Department and an internship at the University Hospital Zagreb, Croatia, as well as volunteer work at the Red Cross. He has published several scientific papers and several book chapters and has actively participated in many international scientific conferences. He has also participated in research projects at pre-clinical and clinical levels, as well as in pharmacological clinical trials as a sub-investigator. He is a guest editor of several special journal issues and an editorial board member for several international scientific journals. Dr. Kibel has edited one book on the renin-angiotensin system and co-edited one book on cardiac diseases.

Dr. Michael S. Firstenberg is a board-certified cardiothoracic surgeon who predominately practices at the Maui Memorial Medical Center, Hawaii, USA. He currently holds appointments in the Colleges of Medicine and Graduate Studies at Northeast Ohio Medical University, USA. He attended Case Western Reserve University Medical School and completed fellowships at The Ohio State University (Thoracic Surgery) and The Cleveland Clinic (Surgical Heart Failure), USA. He is an active member of numerous professional societies and committees, including the American College of Academic International Medicine in which he is a founding fellow and past president. He has authored more than 200 peer-reviewed manuscripts, abstracts, and book chapters. He has edited several textbooks and lectured worldwide on topics such as medical leadership, patient safety, and extracorporeal membrane oxygenation.

Contents

Preface

Diseases of the vascular system represent a global healthcare burden. Acute or chronic occlusive and thromboembolic complications of the arterial and venous systems account for substantial individual and socioeconomic problems at all stages of life. Lifelong disease progression can have devastating impacts on quality of life and life expectancy, as complications of various vascular diseases are leading causes of stroke, heart disease, renal failure, blindness, and limb loss. Such problems are also typically made worse by a synergistic interaction with various morbidities such as diabetes, smoking, obesity, hypertension, and hyperlipidemia. Furthermore, the impact of genetics, diet, and a sedentary lifestyle cannot be overemphasized. While it is well established that such problems at the "large" or macro-circulation level can result in considerable morbidity and mortality, including premature death, limb loss, heart failure, and stroke, the growing impact of diseases of the microcirculation is also evolving. Conceptually, microcirculation represents the bridging vessels between the major branches of the arterial and venous conduits. Typically, these bridging vessels are considered arterioles, venules, and capillaries. Despite their small, microscopic size, their critical role in oxygen and nutrient delivery to the end-organ tissues and as a direct conduit for the removal of biological metabolic waste products of aerobic and anaerobic metabolism demonstrate their importance in preserving organ function and organism survival.

As one starts or continues the journey to better understand the role of microcirculation in normal human physiology and states of disease, it is important to recognize that these structures do not exist and function in isolation separate from the rest of the body – and vice versa. A thromboembolic disease of the larger arteries, for example, can have a substantial impact on the normal structure and function of the microcirculation. Several of the chapters in this book specifically address this delicate balance and relationship. Likewise, while thromboembolic and occlusive diseases of the larger arteries and veins might be better understood, especially considering their potential "downstream" impact on blood flow and oxygen delivery, the impact on the local or regional tissues could potentially be more significant with regards to organ, tissue, and organism structure, function, and viability. Microcirculation occlusive diseases, either acute or chronic, not only have direct effects on distal organ ischemia and metabolism but also have systemic effects from the release of toxic mediators, such as lactic acid from anaerobic metabolism, and paracrine biomarkers or biologically active compounds such as enzymes and cellular receptor activators. Understanding these complex interactions represents mechanistically, as discussed in several chapters, opportunities to improve our abilities to diagnose, manage, and monitor disease progress and the responses to therapeutic interventions. Several of the chapters in this book specifically examine some of the potential pharmacologic tools that can be used to address some very difficult-to-manage clinical problems.

The response of the microcirculation to tissue injury and trauma demonstrates its important role in not only facilitating tissue healing but also maintaining the delicate balance of hemostasis. While the complex interactions that occur because of tissue injury, and the stages of wound healing (both normally and in states of disease), are well beyond the scope of this text, a basic understanding of the interactions between the microcirculation and the blood that is flowing through the body is important. The microcirculation does not serve as a mere passive conduit for blood flow like water through pipes in a house, but rather as an active mediator of the clotting cascade. When a tissue bed is injured or the integrity of the vascular system is disrupted, there are local cellular mediators, such as receptor-activating proteins, that are released that assist in the normal response to tissue injury and subsequent wound healing. Crucial to this process is the simple need, but complex process, to stop the bleeding. Essential in this step is the release of platelet-activating factors that initiate the dynamic steps necessary to induce local thrombosis to "plug a hole." Of course, this process must include a feedback mechanism to ensure the resumption of normal blood flow when the injury is repaired and to prevent excessive clotting that might induce further substantial thromboembolic challenges. Understanding these relationships, cellular communication feedback, and receptor-activating pathways is not only important in trying to manage conditions of tissue injury but also in reducing the impact of abnormalities in these pathways when they do occur. For example, and as discussed in this book, manipulating or reducing the magnitude of platelet activation has been shown to reduce the risk of developing and worsening thromboembolic complications in the setting of occlusive vascular diseases, such as strokes and acute coronary syndromes. This book discusses several specific disease states, as a foundation for further study and understanding, to help better understand the potential opportunities for interventions in diseases, such as lower extremity arterial diseases, with regards to reducing the substantial healthcare burden from their short- and long-term complications. This reflects the delicate balance that has evolved over time between "too much bleeding" and "too much clotting" to find a balance that is "just right" and recognize that there can be circumstances in which tipping that balance in one direction or another might be clinically desirable.

It is also recognized that diseases or abnormal structure and function of the microcirculation impact all stages of life. Typically, the focus is on patients as they age and the impact of modifiable risk factors, such as smoking and diet, and unmodifiable factors, such as genetics, on disease progression and treatment options as they worsen over a lifespan. However, it is also important to understand that even in utero there are pathologic conditions that can have a substantial impact on the newborn child. The placenta represents one of the most important microcirculatory tissue beds concerning the eventual quality of life and life expectancy. The ability of the pregnant mother to pass on to her fetus adequate oxygenation and nutrient support is critical to a normal and healthy gestation. Impairments of the placental microcirculation, specifically at the trophoblast levels, have been directly correlated to a variety of devastating birth complications whose impact on maternal–fetal morbidity and/or mortality can be devastating at the individual, family, community, and global levels. While the magnitudes and opportunities for interventions are difficult to quantify, without a doubt they are substantial and probably poorly understood and underappreciated.

The role of microcirculation in both health and disease is substantial at all levels and, as discussed, represents substantial individual and social burdens concerning costs to healthcare systems and the community as well as the lost quality of life to the individual. Understanding the role of the microcirculation in both hemostasis and homeostasis is vital in the diagnosis and treatment of diseases of the circulatory system whose global impact is probably one of, if not the, greatest cause of premature death and impaired quality of life. In recognizing this, it then becomes easy to appreciate the tremendous resources and efforts that are being devoted to developing tools and techniques to minimize the impact (including economic and lost productivity) of these devastating problems. A comprehensive review of the overall topic is far beyond the scope of this text; nevertheless, as the editors, we hope to inspire readers and students at all levels to use this book as a basic foundation to inspire lifelong study of this ever-growing and evolving constellation of problems. We wish to express our great appreciation to the contributors to this project for their efforts and tireless passion in helping all of us to better treat patients.

Aleksandar Kibel, MD
Department of Cardiology,
University of Osijek,
Osijek, Croatia

Michael S. Firstenberg, MD
Department of Cardiothoracic Surgery,
Maui Memorial Medical Center,
Wailuku, Hawaii, USA

Chapter 1

Structural Characteristic of Growth, Maturation, and Spatial Arrangement of Capillary Bed in Normal and Pathologic Placenta

Marie Jirkovská

Abstract

Placental capillary bed plays a key role in the bidirectional transport between mother and fetus. Its continuous growth and maturation accompany fetal growth and meet all fetal requirements to secure fetal well-being. Considerable growth of both capillary bed and area of villous syncytiotrophoblast comes on in third trimester of pregnancy, continues until the end of pregnancy, and is expressed by rapid development of terminal villi. The presented structural and quantitative data show enhanced villous capillary branching, higher proportion of capillaries displaying delayed maturation, and lower proliferative potential of cells forming capillary wall and cytotrophoblast in diabetic placenta at term. Too few studies have focused on the impact of other pathologies, i.e., preeclampsia and IUGR on development of placental capillary bed. The further research may contribute to better understanding of those disorders connected with pregnancy.

Keywords: capillary, nestin, pericyte, proliferation, spatial arrangement

1. Introduction

The vascular system is inevitable organ component ensuring the supply and outflow of the blood and the blood distribution by arteries, arterioles, venules, and veins, as well as transport of gases, ions, nutrition, wastes, and signaling molecules between blood and cells at the level of capillaries. The development of vascular bed is one of processes of the organ growth, and after the end of its development the vascular component of the organ remains very stable under physiological conditions. Unlike other organs, there are locations in the human body displaying periodical development and demise of vascular bed, i.e., corpus luteum and endometrium.

The third organ, also bound to the fertile period of women life and showing vascular development, is placenta. The whole organ and obviously its vascular bed grow and develop during the whole pregnancy and follow the growth of fetus. Unlike endometrium and corpus luteum, the time of demise of vascular bed is not clear as the fetal well-being depends on the optimal placental function until the moment of birth. Therefore, studies on the placenta at the end of normal non-complicated gestation are performed on the functioning organ.

2. The spatial arrangement of placental capillaries

The capillary bed of the human *placenta villosa* is found in terminal branches of the villous tree, i.e., terminal villi. Terminal villi are covered with trophoblast. It consists of continuous layer of syncytiotrophoblast and underlying cytotrophoblast cells sparsely scattered in placenta at term. Unlike syncytiotrophoblast, cytotrophoblast cells proliferate and their fusion with syncytiotrophoblast ensures the enlargement of its surface area. The stroma of loose connective tissue contains villous capillary bed. The massive development of terminal villi and their capillaries takes place in the last trimester [1]. In placenta at term, terminal villi of the mean caliber 51 μm forms about 39% of the total villous volume and about 46% of the total villous surface area. The estimation of placental capillaries reaches about 12 m^2 of capillary surface [2].

The spatial arrangement of villous capillaries is important for our knowledge of the normal blood flow in the capillary bed and thus for better understanding of its function in transport between mother and fetus. The routine light microscopy shows that voluminous placental capillaries of uneven diameter are found in tight relationship to the trophoblast covering the villus. Thinner parts of syncytiotrophoblast lacking nuclei together with adjacent capillary wall form frequent vasculosyncytial membranes taken as sites of preferential maternofetal transport (**Figure 1**).

Placental capillaries are of somatic type, i.e., the endothelial cells have tight (occluding) junctions on their contacts, the basal lamina is continuous, and perivascular pericytes and their projections surround the endothelial tube (**Figure 2**).

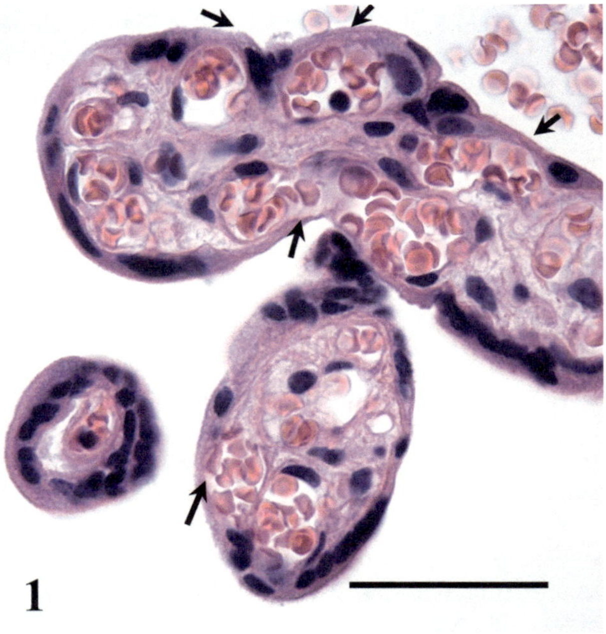

Figure 1.
Histological section showing terminal placental villi. Arrows indicate vasculosyncytial membranes. Scale bar = 50 μm.

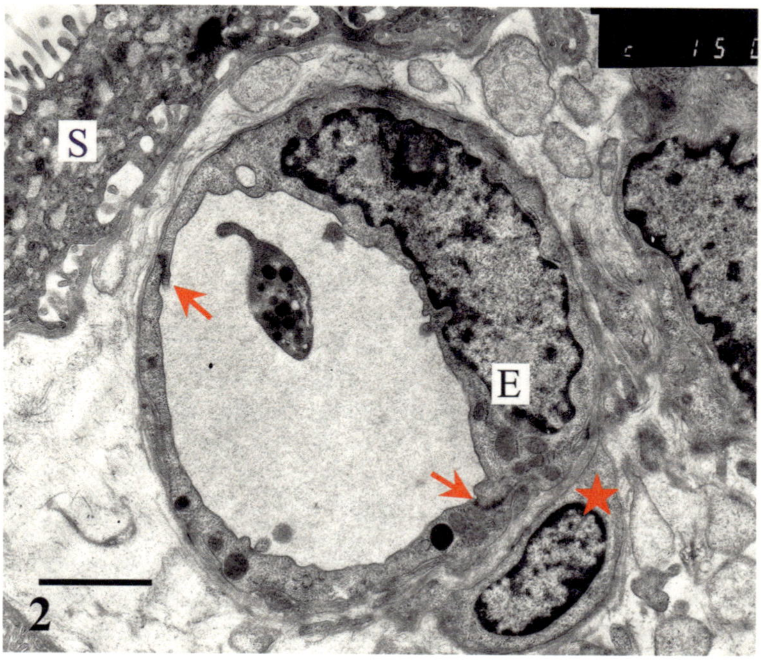

Figure 2.
Placental capillary in a tight relationship with syncytiotrophoblast (S). Endothelial cells (E) are interconnected with occluding junctions (arrows) and surrounded with pericytes (asterisk). Scale bar = 2 μm.

It is hardly possible to judge the spatial arrangement of villous capillaries from conventional two-dimensional pictures. Scarcity of papers dealing with spatial capillary arrangement has shown that it is difficult to achieve reliable data using injection of a dye into placental vessels, corrosion casts, or 3D reconstruction based on serially sectioned material [3, 4]. So, models based on those techniques do not meet the real spatial arrangement of placental capillaries. Only the application of advanced methods using combination of confocal laser scanning microscopy (CLSM) and software for 3D reconstruction yielded successful representation of villous capillary bed and allowed to obtain quite new findings on its real configuration [5].

Terminal villi of cylindrical shape arise from the mature intermediate villi. The simplest type of 3D arrangement is a U-like capillary loop parallel to the villous axis and continuous with capillaries of neighboring terminal villi or arising from a vessel (or also open into a vessel) of mature intermediate villus. It is evident that this type of capillary bed is a result of the longitudinal (nonbranching) capillary growth. The wavy course of some capillary segments represents their continuous elongation inside the villus. In contrast to the previously accepted model [6], the CLSM also revealed that the other type of angiogenesis, the branching angiogenesis is a common process taking part in placental capillary growth as well. It demonstrates itself by the appearance of blind capillary buds, longer blindly ending capillary projections, and of capillary beds formed by three or more capillary segments running parallel to the villous axis. Transversally running capillary segments interconnecting longitudinally oriented capillary segments originate also by the branching angiogenesis and make the capillary bed denser. All capillaries are located in tight relationship to (syncytio)trophoblast, and this way a larger area of vasculosyncytial membranes is formed. Further elongation of both longitudinally oriented capillary segments and transversally oriented capillary segments and blind capillary buds bring about the arching of adjacent trophoblast that is the first visible sign of a new villus formation. As demonstrated here, the continuous elongation and sprouting of capillaries lead to development of villous clusters with an intricately anastomosing capillary bed (**Figures 3–5**).

Concerning the control of villous development, little information is available. Nevertheless, placental vessels belong to the fetal circulatory system, and its capillary part together with trophoblast are fetal structure displaying the anatomically closest relationship to maternal organism and carrying out the transport between mother and fetus. The fetus that is completely dependent on the supply with oxygen and nutrition from the mother reacts on the maternal environment by influencing the growth of placental capillary bed via trophoblastic production of angiogenesis controlling factors [7–9].

As oxygen is essential to fetal survival, its decreased availability elicits hypoxia in fetoplacental unit. One of causes of fetal hypoxia is abnormal glucose metabolism in maternal organism, i.e., insulin-dependent and gestational diabetes. The introduction of insulin allowed the women suffering from diabetes to finish their pregnancy by the birth of viable newborn. But those pregnancies were accompanied with pathological, bulky edematous placentas. Nowadays, the management of maternal diabetes results in consistent metabolic control during pregnancy, well developed newborn, and placentas that do not deviate macroscopically from placentas in uncomplicated pregnancies. Nevertheless, despite the consistent metabolic control, the intermittent excess of glucose load taking place in diabetes has consequences in fetal capillary bed.

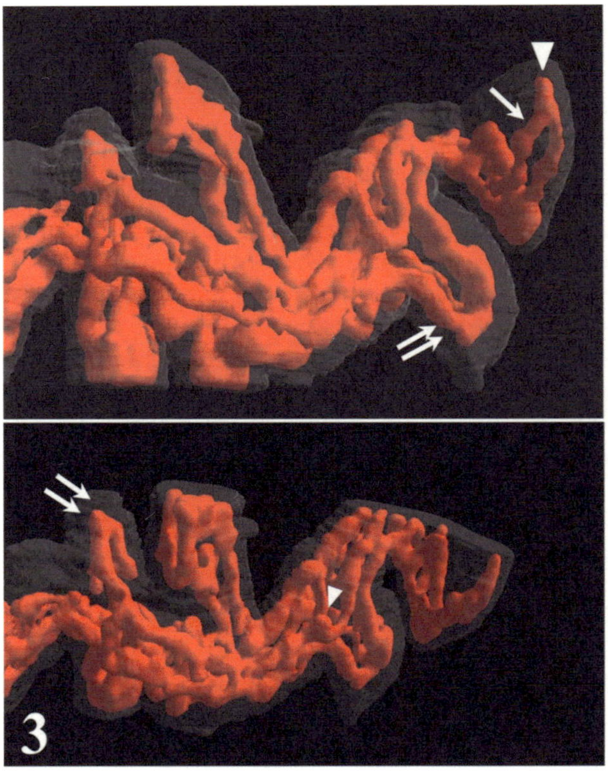

Figure 3.
Two projections of 3D presentation of the villous capillary bed. Blind capillary projections (arrowheads) and villous capillaries formed by three segments (double arrows) illustrate the branching angiogenesis. A simple arrow indicates U-like shape of the villous capillary.

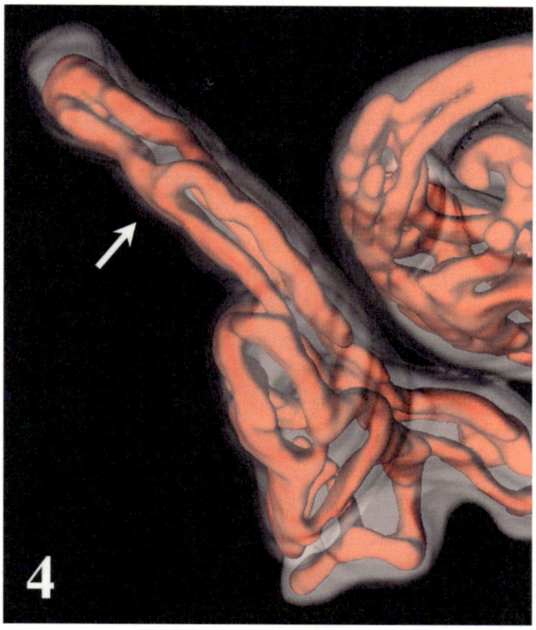

Figure 4.
3D reconstruction of terminal villus with three parallel capillary segments (arrow). Note the apical capillary dilation (arrowhead).

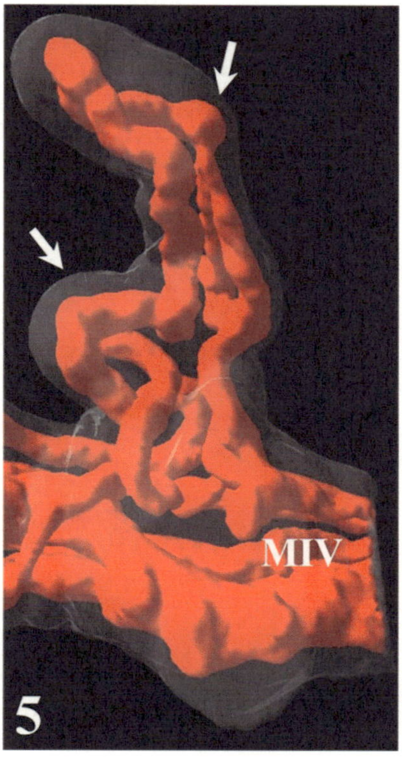

Figure 5.
Developing cluster of terminal villi (arrows) arising from a mature intermediate villus (MIV).

Although nowadays most of placentas from pregnancies complicated by diabetes have histological structure like placentas of non-diabetic mothers, the villous vascularity may be variable, i.e., normally vascularized, edematous hypovascularized villi, and, on the other hand, hypervascularized terminal villi may be found (**Figure 6**).

Quantitative studies based on the confocal microscopic analysis have shown enhanced capillary branching in terminal villi in both gestational and insulin-dependent diabetes [10, 11]. As demonstrated in **Figure 5**, the capillary growth stimulates villous growth and branching, and enhanced angiogenesis in diabetic placenta is probably the cause of higher villous branching described in [12]. Molecular factors conditioning those differences in diabetic placentas were studied, e.g., in [9]. Regarding the vascularization in abnormal villi, three-dimensional presentations show capillary beds formed by thin capillary segments of uniform diameter without dilated parts in hypovascularized villi and, on the other hand, very dilated capillary segments in the capillary bed of hypervascularized villi [3, 11]. The diameter of both forms of those villi are larger (**Figure 6**), and together with more branched villous tree they may disturb the blood flow in the intervillous space and dynamic processes necessary for normal maternofetal transport and capillary blood flow.

The 3D reconstructions of fetal capillaries also bear the potential of better understanding placental physiology and development of placental pathologies. The assessment of the cross-sectional areas of capillaries in capillary bifurcation has shown that villous capillaries branch asymmetrically and that their branching meets the condition for the plasma skimming, the rheological phenomenon important for blood flow in microcirculation [13]. The availability of three-dimensional representations of terminal villi and their capillaries also enabled the computational modeling for simulation and calculation of the influence of pressure drop in capillary dilations on vascular flow resistance and total oxygen transfer rate [14].

Topological characteristics and detailed structural information of villous capillary bed, i.e., length and diameter of capillary segments, their dilation, and branching angle, yielded by three-dimensional imaging, give a data for better insight to blood flow regulation, oxygen transport, blood pressure, and wall shear stress distribution. The further development of simulation methods used in the computational modeling

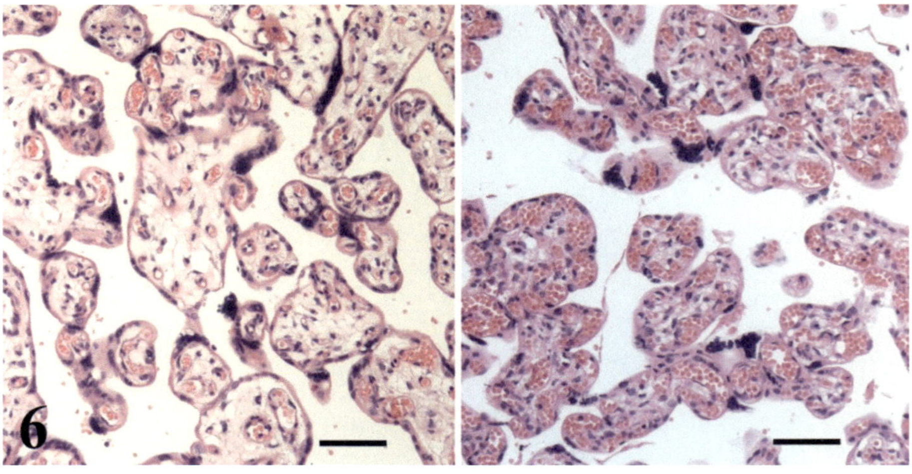

Figure 6.
Hypovascularized (left) and hypervascularized (right) villi of placentas from pregnancies complicated by insulin-dependent diabetes mellitus. Scale bars = 50 μm.

may also contribute to better understanding of dynamic effects of, e.g., capillary diameter and its relationship to blood viscosity and erythrocyte concentration and thus to local differences of blood flow. And obviously, those methods used in the study of pathological placentas may help to elucidate the development and consequences of pathologic villous structure [15, 16].

3. Placental angiogenesis

The development of placental vasculature begins early in the embryonic period from cells of mesodermal origin. Those cells forming mesenchymal core of developing villi give rise to the new vessels in the process of vasculogenesis. Both types of angiogenesis, the nonbranching and branching one, represent the formation of new vessels by elongation or sprouting of preexisting vessels, respectively. Hand in glove with terminal villi, the placental capillaries undergo rapid development during third trimester of pregnancy. Three-dimensional reconstruction of serial confocal sections has shown signs of both nonbranching and sprouting capillary growth taking place inside villi and supporting the growth of new villi [4, 13]. This finding challenges the previously accepted opinion that the nonbranching angiogenesis predominates after 24th week of gestation [1].

The angiogenesis is a process covering the endothelial proliferation, differentiation, and migration allowed by disintegration of endothelial basement membrane. The recruitment of pericytes stabilizes newly formed part of capillary wall or capillary branch, and so, the new structurally and functionally mature part of capillary bed is available.

The cytoskeletal intermediate filament protein nestin occurs transiently among others also in villous capillary wall where it is confined to cytoplasm of endothelium and bodies and projections of pericytes during their differentiation [17–19]. Its immunocytochemical detection shows the distribution of angiogenic "hot spots" in villous capillaries (**Figure 7**). As shown in **Figure 8**, the angiogenic spot may take

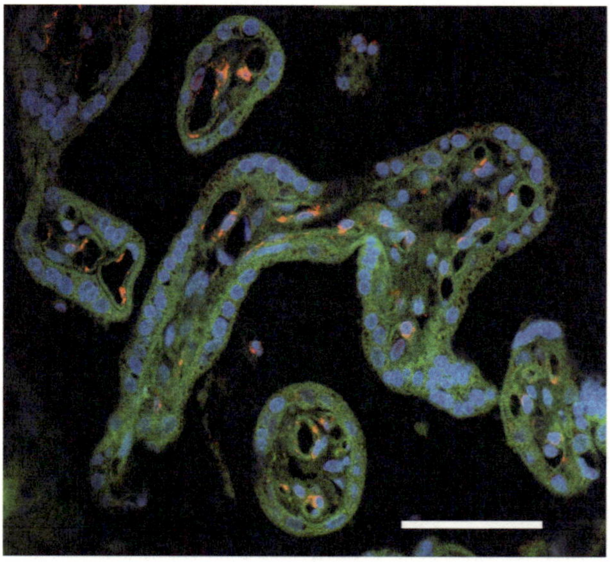

Figure 7.
Distribution of active angiogenic spots (red) in capillaries of terminal villi. Blue signal = cell nuclei (DAPI), green signal = autofluorescence of formalin fixed tissue. Scale bar = 50 µm.

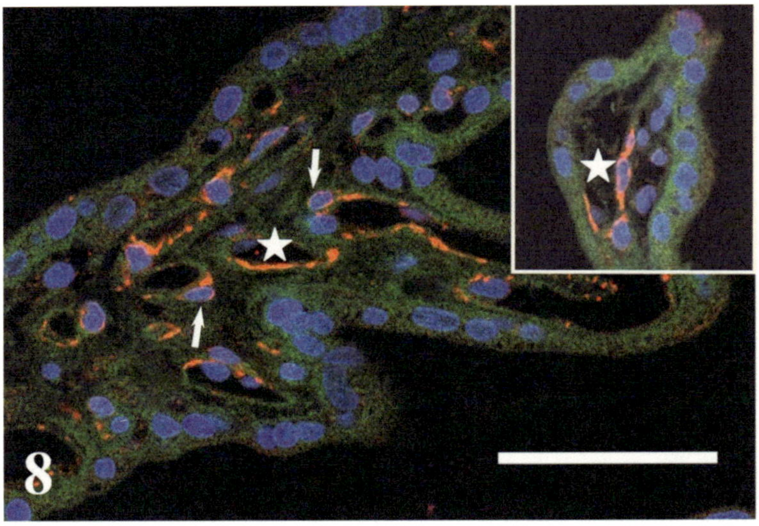

Figure 8.
Immunocytochemical detection of nestin (red) indicates active angiogenic spot of capillary endothelium (asterisks) and differentiating capillary pericytes (arrows). Blue signal = cell nuclei (DAPI), green signal = autofluorescence of formalin fixed tissue. Scale bar = 50 μm.

considerable proportion of capillary circumferrence resulting in enlargement of its diameter. Nestin-labeled cells with protruding nuclei showing typical picture of "ring stone "are differentiating pericytes.

The nestin immunolocalization in the capillary wall reveals characteristic occurrence of nestin-positive endothelium beside quiescent pericytes and nestin-negative endothelium surrounded with nestin-positive pericyte bodies or projections. This picture demonstrates a time shift of differentiation between endothelium and pericytes during formation and maturation of capillaries. The higher proportion of the nestin-positive segments of the capillary circumference in diabetic placentas points stronger angiogenesis in placenta, but it may be also a sign of altered maturation of the cytoskeleton [20].

In semiquantitative study, the authors pointed higher nestin expression as a marker of higher capacity for angiogenesis in preeclamptic placentas [19]. Their finding seems to be in accordance with hyperramification of villous capillaries ascertained in pre-eclampsia using CLSM analysis [21].

In villous capillaries, pericytes are closely associated with endothelial cells and play important role in angiogenesis and vessel stabilization. At present, several new functions in placental development and homeostasis were discovered, i.e., in tissue regenerative and repair processes, lymphocyte activation or phagocytic properties [22].

The arrangement of pericyte coverage and its extent are two features potentially influencing the permeability of villous capillaries. The reduced thickness of tissues insulating maternal and fetal blood that is characteristic for term placenta is expressed most of all in vasculosyncytial membranes (**Figure 1**). As described in [23] and demonstrated in **Figure 9**, pericytes cover mainly endothelial junctions in regions of capillary wall turned away from the trophoblast. That distribution supports maternal-fetal transport and, at the same time, controls intercellular junctions of endothelium. Recently, an advanced method of electron microscopy, i.e., serial block-face scanning electron microscopy, revealed that there are up to 7 μm long interendothelial protrusions [24]. Their coincidence with pericytes maintains without questions the integrity of capillary wall.

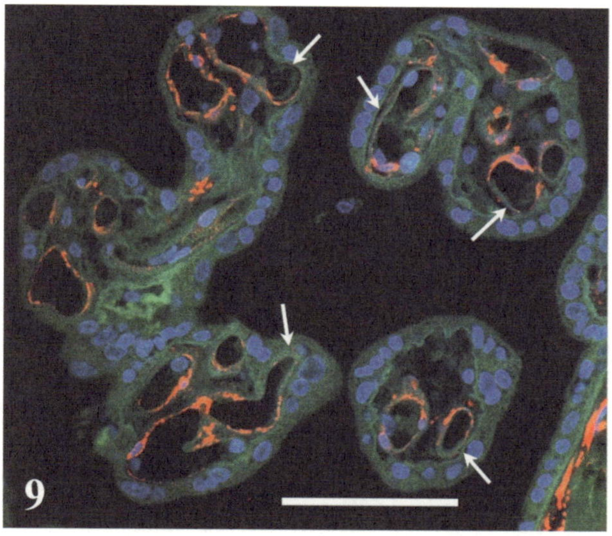

Figure 9.
Immunocytochemical detection of smooth muscle actin labels capillary pericytes (red signal). The pericyte coverage is markedly reduced in regions of capillary wall tightly adjacent to syncytiotrophoblast (arrows). Blue signal = cell nuclei (DAPI), green signal = autofluorescence of formalin fixed tissue. Scale bar = 50 μm.

Quantitative data on the extent of pericyte coverage of capillary endothelium are sparse. The proportion of pericyte coverage was found 15% in normal placenta [24]. The comparison of first trimester placentas and placentas at term has shown higher proportion of vessels surrounded with pericytes at term. Moreover, placental capillaries of pregnancies from high altitude were less covered with pericytes than those from downland [25]. In our study, non-significant differences were found in the proportion of pericyte-associated capillaries in placentas from pregnancies complicated by insulin-dependent diabetes (IDDM) (84% in controls vs. 79% in diabetic placentas) and in extent of pericyte coverage of capillary circumference (38% in controls vs. 33% in diabetic placentas) [26]. The data are hardly comparable among papers, as they were achieved on placental material from different diagnoses by different methodology. On the other hand, the continually lower oxygen pressure may cause lower pericyte coverage in placentas from pregnancies in high altitude. The reduced thickness of vascular wall produced in this case may express an adaptation to preplacental hypoxia caused by lower oxygen tension in high altitude [1]. In our study on diabetic placentas, the lower proportions of pericyte coverage were nonsignificant. Here, we have to take into consideration that the hypoxia in diabetic placenta is a consequence of non-enzymatic hemoglobin glycation leading to decreased release of oxygen from maternal erythrocytes. Diabetic mothers included in this study were well metabolically compensated, and thus, the intrauterine hypoxia, if came about, was low and unlike high altitude pregnancies rather intermittent because of obviously intermittent excess of the glucose load.

The angiogenesis and pruning of capillaries are two sides of formation of capillary bed. In the placenta at term, apoptotic cells were identified in cytotrophoblast and villous stroma including endothelium. The apoptosis was compared in normal and pathological placentas, but results are inconsistent due to studied material and used methodology [20, 27–29].

4. Proliferative potential in term placenta

The enlargement of the area of capillary wall and syncytiotrophoblast is essential for fetal growth that is most rapid in last weeks of pregnancy, and the requirement of optimal placental function overlaps the period of parturition. Thus, the potential of endothelial cells and pericytes to undergo mitotic division is retained in placenta at term. The Ki67 antigen is a proliferation marker expressed in the cell nucleus during the G1, S, G2, and M phase of the cell cycle, but not in quiescent cells. It detects cytotrophoblast, stromal cells, endothelium, and perivascular cells (**Figure 10**). The comparison of normal and diabetic placentas has shown the same occurrence of Ki67 positive cells. But the hypo- and hypervascularized villi of diabetic placentas were discovered free of Ki67-labeled cells [20].

Many studies comparing normal and pathological placentas focus proliferative potential of cytotrophoblast only, e.g. [30]. The cytotrophoblast proliferation potential in peripheral villi was found not different between normal placentas and placentas in intrauterine growth restriction (IUGR), whereas lower proliferation potential was found in stromal compartment in IUGR [31]. The authors of two semiquantitative studies described significantly increased the PCNA and Ki67 signals in the villous compartment of preeclamptic and diabetic placentas [32, 33]. The positive influence of maternal physical activity on increased cell proliferation in both cytotrophoblast and endothelial and stromal cells was proved in [34].

Our analysis performed on terminal villi of normal and DM 1 placentas distinguished between cytotrophoblast, villous stroma, and capillaries. The proliferation index of cells of capillary wall was revealed significantly lower in diabetic placentas. Also, the counts of Ki67-labeled nuclei per villous area unit were significantly lower in cytotrophoblast and capillary wall in terminal villi of diabetic placentas, [20]. Later, the study was extended to the group of placentas from pregnancies complicated by gestational diabetes (GDM) and was also performed in stem and intermediate villi, see **Table 1**.

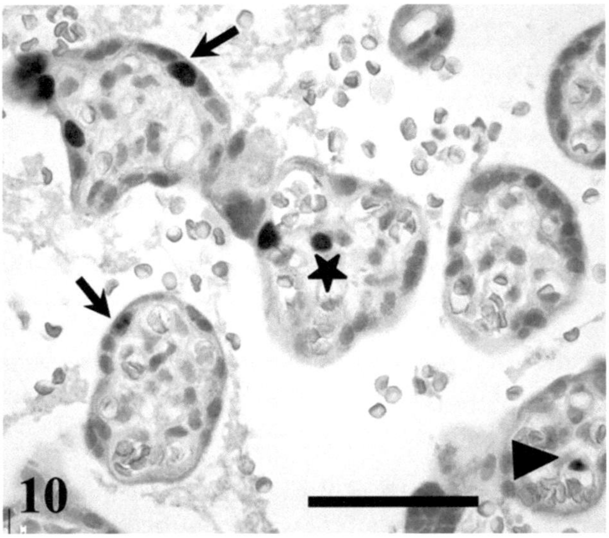

Figure 10.
Ki67-labeled cytotrophoblast (arrows), endothelial cell (arrowhead), and stromal cell (asterisk) in placental villi at term. Scale bar = 50 μm.

Group (number of placentas)	IDDM (16)	GDM (13)	Control (9)
Stem villi[b]			
Cytotrophoblast	6.21 ± 5.36	6.77 ± 5.27	7.60 ± 5.79
Stroma	5.11 ± 6.51	1.68 ± 1.35	2.91 ± 2.47
Vascular wall	1.76 ± 1.94*	1.51 ± 1.53**	3.71 ± 2.07
Intermediate villi[b]			
Cytotrophoblast	9.46 ± 6.12*	10.93 ± 6.42*	18.262 ± 7.741
Stroma	5.43 ± 6.16	1.65 ± 1.13	3.10 ± 4.30
Vascular wall	1.42 ± 1.54	2.07 ± 2.03	4.30 ± 5.51
Terminal villi[a]			
Cytotrophoblast	18.18 ± 11.90*	17.67 ± 9.01*	29.47 ± 15.87
Stroma	8.94 ± 9.78	1.57 ± 1.02	5.42 ± 7.17
Vascular wall	4.48 ± 3.28*	7.39 ± 4.26	9.82 ± 4.46

$^{*}p < 0.05$.
$^{**}p < 0.01$.
[a]Published in [20].
[b]Unpublished data.

Table 1.
Proliferative potential in normal term placentas and placentas from pregnancies complicated by gestational (GDM) and insulin-dependent diabetes (DM 1) was expressed as mean number of Ki67-labeled nuclei per square millimeter of villous cross section.

In stem villi, no significant differences were found out in cytotrophoblast and stromal compartment. The vascular component consists of large arteries and vessels accompanied with capillaries that are taken as *vasa vasorum* of large vessels. The Ki67-labeled cells occurred mainly in their wall, but in lower amount in both IDDM and GDM groups.

Intermediate villi are branches of stem villi characterized by smaller diameter and occurrence of loose bundles of connective tissue fibers in stroma and arterioles, venules, and capillaries. The proliferative potential expressed as count of Ki67-labeled nuclei per villous area unit was discovered lower in cytotrophoblast of both diabetic groups.

The most peripheral branches of the villous tree, terminal villi, massively develop in the third semester to meet growing fetal requirements. Identically to other villous types they are covered with trophoblast, and their sparse stroma contain voluminous capillaries. Cytotrophoblast of those villi displayed significantly lower proliferative potential in both groups of diabetic placentas, whereas the proliferative potential in cells of capillary wall was significantly lower in DM 1 placentas only. Results of that study show that in the placenta at term, there are cell populations, i.e., cytotrophoblast and cells forming capillary wall, available and ready for mitotic division and thus for the enlargement of placental transport area, and that the proliferative potential is reduced in placentas from pregnancies complicated by maternal diabetes.

During their life, normal cells undergo limited number of replications and divisions of chromosomes. This fact limits their lifespan. Among other factors, the replication is regulated by the length of telomeres, and every chromosome replication and mitotic division shortens the telomere length. Telomere shortening induces cell senescence, apoptosis, or genome instability. Some papers showing that the telomere length in placenta decreases during pregnancy were published, but there are no unequivocal results of the assessment of telomere length in pathologic placentas [35]. As those studies were based mainly on the measurement of mean telomere length, the use of contemporary methods quantifying very short telomeres may better clarify processes related to placental senescence at the end of normal and pathologically complicated pregnancies [36, 37]. Nevertheless, it is possible to judge that the shortening of telomeres influences the proliferative potential including term placentas. The more branched villous capillary bed and foci of hypervascularized villi with extremely voluminous capillaries in diabetic placentas [11, 12] represent a compensation of hypoxia. However, such enlargement of capillary wall requires more frequent mitotic division. It is demonstrated by higher proportion of endothelial cells and pericytes labeled with nestin appearing in diabetic placentas. Due to this enhanced mitotic activity, the proliferative potential may be exhausted, and the area of syncytiotrophoblast and capillary wall cannot enlarge adequately to fetal requirements of nutrients and oxygen supply. Such situation might result in placental insufficiency, a serious complication of late pregnancy that is more frequent in maternal diabetes and brings a risk to the fetus.

5. Conclusions

Complications of pregnancy, i.e., maternal diabetes mellitus, preeclampsia, and intrauterine growth restriction threaten the normal intrauterine development of individual. Placenta, the organ ensuring optimal conditions for fetus during pregnancy, mirrors pregnancy pathologies in abnormal structural features. Here, we turned our attention to placental capillary bed. In pregnancies complicated by diabetes mellitus, maternal diabetes is treated nowadays very carefully to achieve consistent metabolic control. Thanks to it placentas look macroscopically as well as microscopically normal. Only the comparison of quantitative data, i.e., villous capillary branching, pericyte coverage, and proliferative potential uncovered differences that are important for our understanding of functional consequences. Unfortunately, similar data for placentas in preeclampsia or IUGR are not available, and those pregnancy complications remain a challenge for further research in this area.

Acknowledgements

This work was supported by a project Cooperation of the Charles University.

Conflict of interest

There are no conflicts of interest.

Author details

Marie Jirkovská
First Faculty of Medicine, Institute of Histology and Embryology, Charles University in Prague, Prague, Czech Republic

*Address all correspondence to: mjirk@lf1.cuni.cz

References

[1] Benirschke K, Burton GJ, Baergen RN. Pathology of the Human Placenta. 6th ed. Berlin Heidelberg: Springer-Verlag; 2012. DOI: 10.1007/978-3-642-23941-0

[2] Benirschke K, Kaufmann P. Pathology of the Human Placenta. 6th ed. New York: Springer-Verlag; 1995

[3] Jirkovská M. The morphology of villous capillary bed in normal and diabetic placenta. In: Zheng J, editor. Recent Advances in Research on the Human Placenta. London, UK: InTech; 2012

[4] Jirkovská M. Spatial arrangement of the microvascular bed in the human placenta. Andrology and Gynecology: Current Research. 2013;**1**:4. DOI: 10.4172/2327-4360.1000110

[5] Jirkovská M, Kubínová L, Krekule I, Hach P. Spatial arrangement of fetal placental capillaries in terminal villi: A study using confocal microscopy. Anatomy and Embryology. 1998;**197**:263-272. DOI: 10.1007/s004290050136

[6] Kaufmann P, Bruns U, Leiser R, Luckhardt M, Winterhager E. The fetal vascularisation of term human placenta villi. II. Intermediate and terminal villi. Anatomy and Embryology. 1985;**173**:203-214

[7] Chen DB, Zheng J. Regulation of placental angiogenesis. Microcirculation. 2014;**21**:15-25. DOI: 10.1111/micc.12093

[8] Loegl J, Nussbaumer E, Hiden U, Majali-Martinez A, Ghaffari-Tabrizi-Wizy N, Cvitic S, et al. Pigment epithelium-derived factor (PEDF): A novel trophoblast-derived factor limiting feto-placental angiogenesis in late pregnancy. Angiogenesis. 2016;**19**:373-388. DOI: 10.1007/s10456-016-9513-x

[9] Loegl J, Nussbaumer E, Cvitic S, Huppertz B, Desoye G, Hidden U. GDM alters paracrine regulation of feto-placental angiogenesis via trophoblast. Laboratory Investigation. 2017;**97**:409-418. DOI: 10.1038/labinvest.2016.149

[10] Jirkovská M, Kubínová L, Janáček J, Moravcová M, Krejčí V, Karen P. Topological properties and spatial organization of villous capillaries in normal and diabetic placentas. Journal of Vascular Research. 2002;**39**:268-278. DOI: 10.1159/000063692

[11] Jirkovská M, Kučera T, Kaláb J, Jadrníček M, Niedobová V, Janáček J, et al. The branching pattern of villous capillaries and structural changes of placental terminal villi in type 1 diabetes mellitus. Placenta. 2012;**33**:343-335. DOI: 10.1016/j.placenta.2012.01.014

[12] Björk O, Persson B. Villous structure in different parts of the cotyledon in placentas of insulin dependent diabetic women. Acta Obstetrica Gynecologica Scandinavica. 1984;**6**:37-43

[13] Jirkovská M, Janáček J, Kaláb J, Kubínová L. Three-dimensional arrangement of the capillary bed and its relationship to microrheology in the terminal villi of normal term placenta. Placenta. 2008;**29**:892-893. DOI: 10.1016/j.placenta.2008.07.004

[14] Pearce P, Brownbill P, Janáček J, Jirkovská M, Kubínová L, Chernyavsky IL, et al. Image-based modeling of blood flow and oxygen transfer in feto-placental capillaries. PLoS One. 2016;**11**:e0165369. DOI: 10.1371/journal.pone.0165369

[15] Plitman Mayo R, Olsthoorn J, Charnock-Jones DS, Burton GJ,

Oyen ML. Computational modeling of the structure-function relationship in human placental terminal villi. Journal of Biomechanics. 2016;**49**:3780-3787. DOI: 10.1016/j.jbiomech.2016.10.001

[16] Plitman MR. Advances in human placental biomechanics. Computational and Structural Biotechnology Journal. 2018;**24**:298-306. DOI: 10.1016/j.csbj.2018.08

[17] Michalczyk K, Ziman M. Nestin structure and predicted function in cellular cytoskeletal organisation. Histology and Histopathology. 2005;**20**:665-671. DOI: 10.14670/HH-20.665

[18] Mokrý J, Němeček S. Angiogenesis of extra-and intraembryonic blood vessels is associated with expression of nestin in endothelial cells. Folia Biologica (Praha). 1998;**44**:155-161

[19] Hwang HS, Cho NH, Maeng YS, Kang MH, Park YW, Kim YH. Differential expression of nestin in normal and pre-eclamptic human placentas. Acta Obstetrica Gynecologica Scandinavica. 2007;**86**:909-914. DOI: 10.1080/00016340701417018

[20] Jirkovská M, Kučera T, Dvořáková V, Jadrníček M, Moravcová M, Žižka Z, et al. Impact of maternal diabetes type 1 on proliferative potential, differentiation, and apoptotic activity in villous capillaries of term placenta. Placenta. 2016;**40**:1-7. DOI: 10.1016/j.placenta.2016.02.003

[21] Resta L, Capobianco C, Marzullo A, Piscitelli D, Sanguedolce F, Schena FP, et al. Confocal laser scanning microscope study of terminal villi vessels in normal term and pre-eclamptic placentas. Placenta. 2006;**27**:735-739. DOI: 10.1016/j.placenta.2005.07.006

[22] Barreto RSN, Romagnolli P, Cereta AD, Coimbra-Campos LMC, Birbrair A, Miglino MA. Pericytes in the placenta: Role in placental development and homeostasis. Advances in Experimental Medicine and Biology. 2019;**1122**:125-151. DOI: 10.1007/978-3-030-11093-2_8

[23] Harris SE, Matthews KSH, Palaiologou E, Tashev SA, Lofthouse EM, Pearson Farr J, et al. Pericytes on placental capillaries in terminal villi preferentially cover endothelial junctions in regions furthest away from the trophoblast. Placenta. 2021;**104**:1-7. DOI: 10.1016/j.placenta.2020.10.032

[24] Palaiologou E, Goggin P, Chatelet DS, et al. Serial block-face scanning electron microscopy reveals novel intercellular connections in human term placental microvasculature. Journal of Anatomy. 2020;**237**:241-249. DOI: 10.1111/joa.13191

[25] Zhang EG, Burton GJ, Smith SK, Charnock-Jones DS. Placental vessel adaptation during gestation and to high altitude: Changes in diameter and perivascular cell coverage. Placenta. 2002;**23**:751-762. DOI: 10.1053/plac.2002.00856

[26] Kučera T, Vyletěl I, Moravcová M, Krejčí V, Žižka Z, Jirkovská M. Pericyte coverage of fetoplacental vessels in pregnancies complicated by type 1 diabetes mellitus. Placenta. 2010;**31**:1120-1122. DOI: 10.1016/j.placenta.2010.09.014

[27] Smith SC, Baker PN, Symonds EM. Increased placental apoptosis in intrauterine growth restriction. American Journal of Obstetrics and Gynecology. 1997;**177**:1395-1401. DOI: 10.1016/s0002-9378(97)70081-4

[28] Sgarbosa F, Barisan LF, Brasil MAM, Costa E, Calderon IMP, Gonçalves CR, et al. Changes in apoptosis and Bcl-2 expression in human hyperglycemic term

placental trophoblast. Diabetes Research and Clinical Practice. 2006;**73**:143-140. DOI: 10.1016/j.diabres.2005.12.014

[29] Burleigh DW, Stewart K, Grindle KM, Kay HH, Golos TG. Influence of maternal diabetes on placental fibroblast growth factor-2 expression, proliferation, and apoptosis. Journal of the Society of Gynecological Investigation. 2004;**11**:36-41. DOI: 10.1016/j.jsgi.2003.06.001

[30] Can M, Guven B, Bektas S, Arikan I. Oxidative stress and apoptosis in preeclampsia. Tissue and Cell. 2014;**46**:477-481. DOI: 10.1016/j.tice.2014.08.004

[31] Chen C-P, Bajoria R, Aplin JD. Decreased vascularization and cell proliferation in placentas of intrauterine growth-restricted fetuses with abnormal umbilical artery flow velocity waveforms. American Journal of Obstetrics and Gynecology. 2002;**187**:764-769. DOI: 10.1067/mob.2002.125243

[32] Unek G, Ozmen A, Mendilcioglu I, Simsek M, Korgun ET. The expression of cell cycle related proteins PCNA, Ki67, p27 and p57 in normal and preeclamptic human placentas. Tissue and Cell. 2014;**46**:198-205. DOI: 10.1016/j.tice.2014.04.003

[33] Unek G, Ozmen A, Mendilcioglu I, Simsek M, Korgun ET. Immunohistochemical distribution of cell cycle proteins p27, p57, cyclin D3, PCNA and Ki67 in normal and diabetic human placentas. Journal of Molecular Histology. 2014;**45**:21-34. DOI: 10.1007/s10735-013-9534-3

[34] Bergmann A, Zygmunt M, Clapp JF. Running throughout pregnancy: Effect on placental villous vascular volume and cell proliferation. Placenta. 2004;**25**:694-698. DOI: 10.1016/j.placenta.2004.02.005

[35] Jirkovská M, Korabečná M, Laššáková S. Telomeres and telomerase activity in the human placenta. In: Morrish TA, editor. Telomerase and non-Telomerase Mechanisms of Telomere Maintenance. London, UK: IntechOpen; 2020. DOI: 10.5772/intechopen

[36] Edelson PK, Sawyer MR, Gray KJ, Cantonwine DE, McElrath TF, Phillippe M. Increase in short telomeres during the third trimester in human placenta. PLoS One. 2022;**17**:e0271415. DOI: 10.1371/journal.pone.0271415

[37] Lai T-P, Simpson M, Patel K, Verhulst S, Noh J, Roche N, et al. Telomeres and replicative cellular aging of the human placenta and chorioamniotic membranes. Scientific Reports. 2021;**11**:5115. DOI: 10.1038/s41598-021-84728-2

Chapter 2

Interplay between Platelet Dysfunction and Vascular Thrombosis in Traumatic Injury

Gordon Ogweno and Edwin Murungi

Abstract

Platelets halt bleeding accompanying traumatic injury by performing primary hemostasis to repair vascular leakage at injury sites. In trauma individuals, *ex vivo* platelet function tests often indicate impairment despite normal count. Moreover, incubation of platelets from normal non-traumatized individuals with plasma from trauma victims demonstrates impairment suggesting association with factors in circulation. Notably, not all trauma victims die from hemorrhage. Despite laboratory evidence of dysfunction, thrombotic vascular occlusions are persistent in trauma survivors as corroborated by postmortem findings from victims who die. The time course of platelet reactions post-traumatic injury, that is, the transition from states favoring bleeding to those that facilitate thrombosis is still unclear. Of the several terminologies describing platelet behavior with regards to injury, including hyporeactivity, anergy, exhaustion, and maladaptive states, few have focused on platelet-platelet interactions. It is increasingly becoming clear that platelet interaction with injured endothelium is a probable missing link in the mechanistic explanation of vascular thrombosis post-traumatic injury. This postulate is supported by evidence of increased adhesive protein, von Willebrand factor, and released from injured endothelium. In all, this potentially explains the suboptimal response to anticoagulants or antiplatelets post-trauma. This chapter will review current knowledge on platelet functions in relation to vascular thrombosis post-trauma, the time course, mechanistic hypothesis, and response to therapeutic interventions and clinical outcomes.

Keywords: traumatic injury, platelet dysfunction, vascular thrombosis, therapeutic interventions, interplay

1. Introduction

1.1 Clinical presentation of platelet dysfunction

Platelet dysfunction post-trauma presents as excessive and immediate bleeding in distinct patterns including scattered small ecchymoses at sites of minor trauma or venipuncture, spontaneous bleeding at various body surfaces such as mucosal (oropharyngeal, genitourinary, gastrointestinal, nasal), ecchymosis, petechiae (mostly evident on lower limbs), purpura, epistaxis, and gingival bleeding. Bleeding into deep tissues, joints, or hematomas formation is rare [1].

Alternatively, hyperactive platelets contribute to occlusive vascular thrombus formation. Although thrombus formation on the venous system is generally rare, arterial thrombus on vascular territories associated with myocardial infarction and ischaemic stroke are common.

1.2 Platelet morphology and structure

Platelets are anucleate smallest blood cells with an average lifespan of 10 days. In circulation at rest they are biconvex, but *ex vivo* in EDTA they appear round on average 2.5 μm diameter and mean platelet volume (MPV) of 8–10 fL but could be as large as 4–5 μm. Their plasma membrane is made up of proteins and lipids (mainly phospholipids and cholesterol). Interspersed within the lipid bilayer are glycoproteins that serve as structural support, receptors for ligands, and components of platelet reactions. Additionally, the plasma membrane extends internally forming membrane-bound open canalicular structures (OCS) that are distinct from dense tubular system (DTS), which are remnants of rough endoplasmic reticulum parallel to the surface with blind endings [2].

Inside platelets are cytoskeletal structures and granular elements. The platelet cytoskeletons are made up of microtubules and microfilaments that are responsible for maintaining discoid shape and shape change, extension of pseudopods, and secretion of granule contents. The cytoplasm contains several granules named according to appearance on electron microscopy. Dense granules are electron-dense and contain simple molecules such as Ca^{++}, Mg^{++}, ATP, ADP, and serotonin. The less electron opaque alpha granules mostly contain proteins including vWF, fibrinogen, thrombospondin, coagulation factor V (FV), β-thromboglobulin (β-TG), P-selectin, glycoproteins (GP), HMWK, plasminogen, plasmin and inhibitors, and fibronectin [2]. Upon platelet activation, the granular contents are released into the canalicular system to participate in platelet reactions.

1.3 Platelet reactions at injury site

Under normal physiological conditions, circulating platelets are non-thrombogenic, neither adhering to endothelium nor aggregating to each other. This is because normal endothelial lining constantly releases suppressing factors such as nitric oxide (NO), prostacyclin (PI), and an ADPase (CD39) [3, 4]. Injury activates the endothelium to release ADP, thromboxane, and prostaglandins and promotes thrombin generation. Moreover, injury exposes subendothelial matrix composed of adhesive proteins collagen and vWF to circulating blood. Collectively, these factors lead to cascade of reactions on platelets including adhesion, shape change and spreading, granule secretion, thromboxane synthesis, and aggregation (**Table 1**) [4].

Following receptor binding, a series of intracellular events ensue that lead to activation of phospholipases and increase in second messengers, synthesis of eicosanoids that initiate further amplification of cellular activities such as shape change, granule secretion, GPIIbIIIa conformational change, membrane reorganization to exteriorize phosphatidylserine (PS), membrane blebbing, and microvesicle/microparticle formation [5, 6]. All these processes of platelet reactions are accompanied by increase in intracellular calcium [4].

Platelet reactions eventually end up in platelet plugs or pathological thrombus formation. Emerging evidence indicates that thrombus components and the signaling

Group	Ligand	Physiological Receptor	Family of receptor	Outcome of ligand-receptor interaction
Adhesive proteins	collagen	GPVI	Glycoproteins (GP) Integrins	Adhesion
	vWF	GPIb-v-IX (GPIbα)		Tethering
	Fibrinogen/fibrin	GPIIbIIIa (PAC-1)		Aggregation
	Immunoglobulins	FcRy& FcyRIIa	ITAM	Activation, aggregation
	Thrombospondin	GPIa-IIa	Glycoprotein	Spreading
Soluble agonists	Adenosine diphosphate(ADP)	P2Y12	G-protein coupled receptor (GPCR)	Activation
	Adrenaline	α2A		Activation
	Thromboxane	TP		Activation
	Arachidonic acid (AA)	PGE/PGI		Activation
	Thrombin	Protease-activated Receptors 1–4 (PAR 1–4)		Activation
Coagulation factors	Thrombinase	PS	Membrane lipid	Assembly of coagulation factors and amplification of coagulation reactions
Alarmins	DAMPS including NETs	Toll-like receptors (TLR)	Lipids-Pattern Recognition Receptors of Pathogen Associated Molecular Patterns(PAMP)	Activation

Table 1.
Platelet ligands in trauma, receptors, and outcome of their interaction.

pathways in platelet plugs are hierarchical. The outer layer of platelet aggregates is composed of loosely packed platelets poor in p-selectin and little and no fibrin. Thus, the decreasing density of fibrin from the core outward parallels thrombin levels, and reflects a hierarchy in the impact of agonists based on composition [7].

To limit thrombus growth and maintain vascular luminal patency, various intrinsic mechanisms that negatively regulate platelet activation come into play including immunoreceptor tyrosine-based inhibition motif (ITIM), endothelial cell-selective adhesion Molecule (ESAM), Wnt-β-catenin, and semaphoring 3A (Sema3A). Furthermore, integrin ectodomain receptors are shed by proteases such as thrombin and ADAM, microvesiculation, and internalization, resulting in loss of adhesive and aggregation features in thrombus formation [8]. These negative regulators limit the intracellular signaling, integrin activation, receptor desensitization, and response to secondary mediators [4]. Therefore, following trauma, platelets undergo changes from quiescence, activation, thrombus growth, and finally to self-regulation [4].

2. Assessment of platelet functions in trauma

Although multiple events occur concurrently in platelets during trauma, laboratory investigations only focus on one or two. Tests performed include (i) routine platelet count and bleeding time, (ii) flow cytometry assays of surface membrane receptor expression, (iii) perfusion analysis of adhesion to collagen or fibrinogen coated surfaces, (iv) analysis of agonist-induced platelet-platelet aggregation by light transmission aggregometry (LTA), impedance aggregometry or platelet function analyzer-100 (PFA-100), or (V) TEG-platelet mapping/ROTEM [9].

2.1 Bleeding time

This was the initial test for primary hemostasis for decoding platelet functions *in vivo*. This test reveals persistent longer bleeding time in traumatic brain injury patients compared to healthy controls patients [10]. However, this test has largely been abandoned due to the lack of specificity and sensitivity and not routinely performed.

2.2 Platelet count

Majority of patients arrive in emergency departments with normal or near-normal platelet counts [11]. In a cohort of trauma patients with mean injury severity score of 22, platelet count progressively dropped over a 72-hour observation period even though it did not reach the critical level associated with spontaneous bleeding [12, 13]. At admission, bleeding and requirements for transfusion occur at much higher platelet counts compared to other conditions [11] suggestive of dysfunction.

2.3 Adhesion

In trauma patients, platelet adhesion, as measured on collagen and fibrinogen by flow chambers, was decreased compared to normal healthy individuals [14].

2.4 Aggregometry

Upon activation, platelets undergo shape change and stick to each other through fibrin bridges. These biophysical changes can be evaluated to ascertain platelet function.

2.4.1 Light transmission aggregometry (LTA)

Considered the gold standard, LTA is a widely used *ex vivo* assay for platelet function [9, 15]. The method has been used to demonstrate a decrease in platelet function in response to ADP and TRAP agonists in trauma patients, findings that correlated with injury severity and level of consciousness but not measures of shock [16]. These results were consistent with those obtained using a modified LTA (optimal) [14]. Despite its centrality, no large-scale randomized studies of LTA in trauma have been conducted.

2.4.2 PFA-100/200™

This test estimates both platelet aggregation and adhesion and has replaced the bleeding time [9]. It has been used to demonstrate platelets dysfunction in a cohort of trauma patients monitored over time. A nested control analysis revealed injured patients with trauma-induced coagulopathy (TIC) patients have longer PFA-100 Coll/Epi and coll/ADP closure time compared (CT) to their non-TIC injured counterparts [17].

In contrast, another study of a cohort of trauma patients showed a shorter PFA-100 closure time at admission compared to controls even though the CT progressively increased returning to normal baseline at 72 hours. Furthermore, closure times were longer in non-survivors compared to survivors [13]. PFA-100 closure time is influenced by a number of factors not specific to platelet functions that include platelet count, RBC, and vWF [18].

2.4.3 Multiple electrode aggregometry (MEA)/impedance aggregometry

This method uses whole blood instead and follows change in electrical resistance between two electrodes as platelets aggregate [19]. It has the advantage of not requiring centrifugation, uses small sample volumes, and is near physiological conditions since platelets are evaluated in the presence of red blood cells and leukocytes similar to *in vivo*.

Most trauma patients arrive at emergency departments with MEA below normal response to ADP, AA, thrombin, and TRAP agonists [20–22], with survivors having less impairment compared to non-survivors [23] and hyporeactivity persisting for 96 hours [24, 25]. Since MEA is sensitive to GP1b deficiency, reduction in total vWF, as well as activity that includes FVIII carrying capacity, the finding of decreased platelet ristocetin response in trauma patients [21, 26] strongly suggests impaired adhesive interaction with endothelium.

2.4.4 VerifyNow™

This is a turbidometric-based optical method detection system, specifically developed to detect sensitivity to antiplatelet therapy. It measures platelet binding to fibrinogen-coated polystyrene beads in whole blood following activation by a number of agonists acting on platelet GPIIbIIIa. Trauma patients not on aspirin have been shown to have greater VerifyNow aspirin reactivity units (ARU) and platelet reactivity units (PRU) compared to those on aspirin [27]. Platelet dysfunction results obtained using this system concurred with those obtained previously in trauma [28] and with those with intracranial hemorrhage [29]. However, a study of platelet function in traumatic injury found a high prevalence of poor platelet response that neither correlated with hemorrhagic outcome nor whole blood aggregometry [30].

3. Thromboelastography (TEG)/thromboelastometry (ROTEM)

A study that evaluated platelet functions using TEG platelet mapping in traumatic brain injury patients revealed decreased response of platelets to AA agonist, more pronounced in bleeders compared to non-bleeders, but no significant differences in ADP stimulation [10]. In another study undertaken in patients with blunt trauma, TEG MA remained unaltered, though MA-Platelet mapping (TEG-PM) AA and ADP were reduced [31]. The platelet inhibition evaluated with TEG occurs early (before 6 hours) post-trauma, worsened with severity of injury [31], hemorrhagic shock, and acidosis [32]. Similar to findings reported using ROTEM-fibtem [14].

A modification of TEG functional fibrinogen level (FLEV-TEG) uses GPIIbIIIa blockers to disentangle fibrin and platelet contributions to clot strength demonstrated that platelet contribution to clot strength at admission accounts for 80% but progressively decreases to 50% over 72 hours then stabilizes for the next 48 hours indicating platelet dysfunction [33], though there are differences in GPIIbIIIa blockers [34, 35].

3.1 Flow cytometry of platelet activation biomarkers

3.1.1 PAC-1 (GPIIbIIIa)

In severely injured patients with injury severity score of 22, admission PAC-1 was ten times higher than controls, progressively decreasing over 72 hours but remaining higher than controls at all the time points [13]. This is in contrast to another study where PAC-1 levels were on an upward trend for both TIC and non-TIC patients [17]. This was in contrast to Verni and coworkers [36] who reported decreased levels in response to ADP and CVX agonists. In this case, the response levels were dependent on calcium concentration.

3.1.2 P-selectin (CD62P)

P-selectins are stored in platelet alpha granules but are translocated to the membrane surface upon activation. There is a direct and linear relationship between increase in P-selectin and clot forming potential as represented by aggregation [37]. On admission, trauma patients have higher platelet P-selectin that reduces over 72 hours but remains above that of controls throughout [13, 17]. This finding contradicts that of Mathay and co-workers who reported low levels at admission [38]. Differences in response to various agonists have been noted: response to CRP-XL, and ADP is greater in healthy individuals than in trauma patients [14, 36]. There are also differences between survivors and non-survivors [17] indicating differences in signaling mechanisms.

Determination of platelet surface expression of P-selectin may not be an accurate measure of platelet prior exposure/activation *in vivo* as they get detached and released into plasma, while the degranulated platelet continues to circulate [39].

3.1.3 Phosphatidylserine (PS)

Trauma platelets showed decreased PS expression in response to ADP and convulxin [36].

3.1.4 CD 40

Traumatic injury is associated with increased expression of CD 40 receptors on platelets, and these interact with ligands on endothelium and leukocytes [40].

3.2 ELISA evaluation of activation dependent soluble plasma biomarkers

Platelet expressions of surface biomarkers are dynamic transient, and over time are shed off into circulating plasma [41]. These include soluble P-selectin (sP-selectin), glycocalcin (soluble form of GPIbα), soluble GP VI (sGPVi), soluble CD40L (sCD40L), metabolic products such as thromboxane (TXA2), thromboglobulin (TBG), and platelet factor 4 (PF4) [42].

Trauma platelets GP VI and GP1b surface expression were less than healthy controls, but paradoxically the soluble plasma concentrations of sGPVI and sGP1b were higher than in controls [14] suggesting increased protease cleavage.

There is some confusion about the physiological dynamics response to most agonists such as thrombin, convulxin, and TRAP/CRP. Unlike GPIIbIIIa, *ex vivo* agonist stimulation leads to decreased surface expression and triggers internalization and vulnerability to proteolytic cleavage into the surrounding medium [43]. In trauma, the soluble form (sGPVI) is increased and correlates with soluble fibrin formation, D-dimers, and development of thrombosis [44].

Soluble CD40 was found elevated in trauma and correlated with endothelial and tissue damage, DAMPs, fibrinolysis, thrombin generation, acidosis, and sympatho-adrenal hyperactivation [45]. Due to the relation with fibrinolysis (D-dimers) and thrombin generation (PF 1.2 and TAT), this could be a reflection of cleavage after surface expression.

4. Secretions

P-selectin is stored in platelet alpha granules and is translocated to the surface upon stimulation and its expression in response to agonist has been used as a marker of secretion. Although trauma patients platelet's p-selectin expression is higher compared to that of healthy controls, it is further elevated in response to agonists [14].

5. Microparticles

On admission, platelet microparticles in trauma patients are usually twice those of controls and remain unchanged for over 72 hours. Interestingly, non-survivors and head-injured patients have high initial microparticle counts but levelsdecrease in 24 hours to approximate that of survivors [13]. The high levels of circulating microparticles in trauma contribute to increased platelet activation [46]. Also, increased platelet microparticles in trauma patients that persist for over 72 hours are implicated in hypercoagulability [47]. On the other hand, low levels of platelet microparticles are associated with bleeding and mortality [48].

In laboratory animals, traumatic injury with shock is accompanied by increased elaboration of platelet microparticles, and these are associated with increased thrombin generation and DVT in mice [49].

6. Imaging/microscopy

A unique phenotype of platelets in trauma has been visualized characterized by transformation into balloon-like structures has been visualized [50]. Ballooning increases membrane surface area for PS exposure and procoagulant thrombus reactions, as well as microvesiculation [51, 52].

7. Spectrum of platelet dysfunction in trauma

7.1 Hyperreactivity

Few studies have reported findings of platelet hyperreactivity or increased response to stimulating agonists in trauma [13]. Platelets are more hyperreactive as demonstrated by amplified binding to fibrinogen in the presence of increased doses of ADP [53]. This phase is immediate within minutes to hours [54] and is often missed by most studies due to timing of blood sampling.

7.2 Hyporeactivity

Despite normal platelet count, below normal or decreased response to stimulating agonists *ex vivo* has been reported in 45% of patients for atleast for one agonist, 91% during intensive care stay [22, 25]. The independent predictors are injury severity, Glasgow Coma Scale and acidosis [24]. However, minor injury has been associated with platelet hypofunction [31]. Moreover, diminished platelet response has been linked to lowered calcium levels [36].

7.3 Platelet granule exhaustion

Despite normal platelet count at admission [11], there is a discordant reduction in aggregation response to stimulating agonists even with increased receptor expression [55]. Many studies of platelet function in trauma have referred to this phenomenon as –'platelet granule exhaustion' [32]' [31] in line with previous findings in other conditions [56–58] that share similarities with storage pool disorders [59]. However, this position lacks consistent support and has been refuted [60] since trauma platelets still retain response to P-selectin expression though with differences in agonists [14] indicating differences in signaling mechanisms rather than exhaustion. Moreover, there is overreliance of platelet aggregation studies which have been shown to be unreliable in storage pool disorders [61, 62].

Due to the inconsistency in terminology such as 'platelet exhaustion' or 'desensitized, stunned, inactive, post activated, dysfunctional, or degranulated' used to describe the platelet dysfunction in trauma. Perhaps better terminologies such as 'functional anergy' [63] and 'agonist refractoriness' [54] are more apt. It has been opined that what is called maladaptive or dysfunctional platelet in the acute phase of trauma is likely a misnomer, and perhaps it could be an adaptive natural selection mechanism for ensuring survival through possible microvascular thrombosis during the low flow states that kick in following hemorrhage in order to maintain organ perfusion [64].

7.4 Mechanisms for hypofunction

Empirical data have shown that the loss of platelet aggregation functions is plausibly due to: (i) loss of adhesive receptors through microvesiculation, downregulation/internalization [65], and ectodomain shedding of adhesive receptors GP1bα and GPVI [64], (ii) endothelial dysfunction (endotheliopathy) [66] in which glycocalyx release of mediators such as versacan been demonstrated to have impact on platelet dysfunction [67], (iii) reduction in calcium availability [38] and intracellular mobilization [36], (iv) shock acidosis [68], (v) reduced adhesive form of vWF due to hyperactive ADAMTs 13 proteolysis [69, 70], and (vi) low fibrinogen from consumption or defective activity [71].

8. Modifiers of platelet functions in trauma

8.1 Type and severity of injury

The extent of platelet dysfunction worsens with injury severity, acidosis [32], and brain injury [13]. However, platelets functions may be impaired even with minor injuries without acidosis or shock [31], or only correlated with extent of cerebral fatality but independent of injury severity [20].

8.2 Extent of endothelial injury and vWF-ADAMTS-13 axis

An imbalance of vWF: ADAMTS-13 ratio has been found in trauma patients soon after injury and was associated with increased thrombin generation [72]. Low plasma levels of ADAMTS-13 and high vWF are associated with mortality [73, 74], and persistently elevated levels in trauma patients were associated with development of ARDS predictive of survivors from non-survivors [75] indicating link with microvascular thrombosis. Although the increased vWF multimers would be expected to compensate for adhesion of low platelet count in trauma, however, the transient and decreased platelet adhesion and aggregation early in trauma could be multifactorial that include: abnormalities in vWF conformation [26], downregulation of platelet GP 1b receptors [76] from increased thrombin generated [43], and receptor loss through sheddases [8, 14].

8.3 Thrombin generation

There is increased thrombin generation in trauma [77, 78] linked to NETosis and glycocalyx syndecan-1 release [79]. The consequences of increased thrombin generation are platelet receptor activation and fibrin formation that promote aggregation.

8.4 Plasma calcium levels

A study that factored in calcium levels, it was found that platelet activation, aggregation, and membrane surface receptor expression were increased with increasing upward calcium titration [38] indicating importance of calcium-mediated processes.

8.5 Fibrinogen-fibrinolysis axis and role of plasmin

Despite increased expression of platelet GPIIbIIIa aggregation receptors post-trauma [13], most studies report paradoxical reduction in aggregation to most agonists [80]. The time course of reduced platelet aggregation and functional recovery parallels periods of fibrinolysis [81]. The role of plasmin on platelet function is controversial [82], depended on the methodology and testing conditions [83]. Although it has been reported that fibrin proteolytic products mediate platelet dysfunctions [84], it is plasmin that reduces platelet aggregation [85] without affecting GP receptor expression [86]. While the FDPs compete for fibrinogen binding sites on GPIIbIIIa [87] and association with PS- expressed on activated platelets [88], plasmin degrades fibrin/fibrinogen reducing its bridging function between adjacent platelet GPIIbIIIa [89]. Additionally, by cleaving vWF, plasmin reduces platelet adhesion to endothelium [90]. This explains the reduction in ristocetin agglutination in trauma platelets. However, this effect plays a minor role since plasmin also cleaves the regulatory enzyme ADAMTS 13 [91].

On the other hand, plasmin acts as a platelet activator of surface receptor expression and granule secretion under conditions likely found in trauma [92]. The platelet-activating effects of plasmin become evident during fibrinolysis shutdown since only free circulating plasmin are inhibited by plasma antiplasmin and α2-macrogolbulin, as well as PAI-1 on t-PA without affecting platelet bound plasminogen-plasmin. Perhaps the restoration of platelet aggregation by 72 hours post-trauma [93] may be explained by the fibrinolytic shutdown that also occurs at the same time period [94]. The transition from decreased aggregation to restoration and enhanced aggregation could be accounted for by the slow platelet release of PAI-1 [81, 95] that lags behind the plasmin activation but eventually shuts it down [94].

8.6 Damage associated molecular patterns (DAMPs)

Tissue injury, ischemia, and cell death trigger release into plasma damage-associated molecular patterns (DAMPs) including nucleic acids, csDNA, histones, high-mobility group box-1(HMGB-1), heat shock proteins (HSP), and S100 proteins among others [96]. DAMPS also termed alarmins [97] are elevated after trauma [96, 98] and are recognized by toll-like receptors (TLR) on platelets to trigger activation. The time course for plasma DAMPS parallels the duration of platelet hypofunction and recovery [99] strongly suggesting that they are potential drivers of platelet functional fluctuations in trauma in concert with cytokines and fibrinolytic system.

8.7 Neutrophil extracellular traps (NETosis)

Trauma increases platelet P-selectin expression and elevates the levels of neutrophils that release Neutrophil Extracellular Traps (NETs) [100]. NETs, composed of DNA, histones, and neutrophil elastase (NE) in turn promote platelet activation, aggregation, thrombin generation, and thrombosis [101]. In addition, histones promote platelet ballooning and microparticle formation [50] further escalating the risk of vascular thrombosis risks. Thus, the high NETs produced in association with trauma [79] could be considered sentinel markers of platelet activation and thrombosis.

8.8 Shock and acidosis

A study conducted with MEA revealed that injured patients with shock had decreased AUC irrespective of injury severity [68, 102]. During shock states, metabolites together with attendant acids are involved in fibrinolysis that decreases platelet aggregation [103].

Ex vivo, addition of lactic acid to canine platelets to create academia revealed reduced platelet aggregation on MEA [104], consistent with previous studies on LTA [105, 106]. The lowered platelet aggregation due to induced acidosis is linked to modification of intracellular store calcium traffic [107] and inhibition of GPIIbIIIa conformation (PAC-1) [108].

8.9 Inflammation

The intense inflammatory response immediately post-trauma [109, 110] has implications on platelet functions [111]. Interleukins, act directly as potent platelet activators through IL-6 [112, 113], and indirectly through thrombopoietin bone marrow megakaryopoiesis, vWF endothelial release, and sensitization to thrombin [114].

The effects of complements on platelet aggregation in normal and trauma patients are complex and controversial, perhaps reflecting differences in calcium fluxes [115].

8.10 Neurohumoral hormonal axis-sympathoadrenal activation

Trauma is associated with activation of sympathoadrenal system releasing adrenaline and noradrenaline into the circulation [116]. These catecholamines modulate platelet functions indirectly through endothelial damage termed 'endotheliopathy' [66] *via* vWF and exposure of subendothelial collagen, and directly on platelet receptors to cause activation, secretion, and aggregation [117]. Although *in vitro* catecholamines cause platelet hyper aggregation, at the circulating levels in trauma, platelets *ex vivo* response to stimulating agonists is paradoxical decreased aggregation despite increased expression of surface receptors such as GPIIbIIIa, PS, and P-selectin. This phenomenon has variably been referred to as 'granule exhaustion' or 'anergy'. But these references have not taken into account the various substances that are also released with tissue trauma [118].

8.11 Alcohol and toxins

Alcohol is known to have *in vivo* and *ex vivo* platelet aggregation inhibitory effects [119–121], related to alteration of blood osmolality, lactate, and acidosis [122]. Conversely, bidirectional effects termed rebound effects [121] composed of initial decreased aggregation followed by hyperaggregation have been shown [123]. The mechanism for increased aggregation appears to be erythrocyte release of ADP [124] and fibrinolytic shutdown [125, 126]. The bidirectional effects of alcohol may be to the early rise of acidic metabolites such as acetate in the initial phases promoting fibrinolysis, with the late inhibitory effects due to the slow release of PAI-1 that causes *in vivo* shutdown [127].

9. Time course of platelet function in trauma

Early in trauma, there follows a period of acute reduction in platelet aggregation that reaches a nadir after 4–12 hours [25, 36, 128] and gradually returns to normal by 48–96 hours [128]. While the changes in platelet functions are evident as early as one and a half minutes after injury [129], hypofunction may persists for 96 hours [24]. Platelet count also follows similar pattern albeit slowly [80] though onset of recovery occurs earlier than platelet functions [128] that correlates with changes in ADAMTS-13 levels [130]. The trend of initial hypofunction followed by restoration and rebound hyperfunction has been observed in diverse conditions such as head injury [93, 121, 131], critical care units [24], and spinal cord injury [132]. In an animal model of TBI, the changes paralleled increase in pro inflammatory cytokines such as IL-6, KC (keratinocyte chemoattractant), and soluble p-selectin [133]. The observed restoration of platelet functions is consistent with the development of thrombosis [134].

10. Platelet functions and clinical outcomes in trauma

10.1 Bleeding

In the early period following trauma, bleeding is experienced despite normal platelet count [135], a concept referred to as trauma-induced coagulopathy (TIC) [67].

Also, decline in post-traumatic platelet count is associated with intracranial hemorrhage [136]. It is still not clear whether the bleeding is a result of isolated platelet dysfunction, fibrinolysis, or combined effects.

10.2 Vascular thrombosis

Following traumatic injury has been associated with the development of venous thromboembolism in up to 58% of victims [137], although the incidence varies with time [138]. Despite platelet hyporeactivity in the initial phase, restoration of platelet function may trigger rebound hyperaggregation and hypercoagulation [133] potentially leading to development of DVT [22, 25, 139]. Although many trauma patients receive prophylactic anticoagulants, vascular thrombosis still occurs [35].

11. Therapeutic interventions

11.1 Platelet transfusion

Whilst circulating platelets are dysfunctional after trauma [140], platelet transfusion is not associated with restoration of function [54, 128, 141, 142]. The improvement in hemostasis reported in some studies [143] may be attributed to inhibition of fibrinolysis in bleeding trauma patients, but platelets functions remain unaltered compared to those who do not receive transfusion [144]. Notably, platelet transfusion reverses aspirin-induced hypofunction but not trauma-induced dysfunction assessed by MEA [145]. The critical time period when platelet transfusion may be useful has been identified as late in the phase (after 24 h) rather than early (prior to 12 hours) [80]. This time period coincides with decline in aggravating factors such as fibrinolysis, DAMPS, and acidosis allowing restoration of functions to normal levels.

11.2 Antiplatelets

Although some studies have indicated no difference in VTE incidence in surgical patients given antiplatelets [146], data on trauma patients are variable. The incidence of VTE in trauma patients on aspirin prior to injury is half that of matched controls with VTE but not on aspirin [147] suggestive of a protective role. The protective effect of aspirin is enhanced when combined with clopidogrel and systemic anticoagulants such as LMWH or heparin.

11.3 Novel therapeutic targets

In general, therapeutic targets are aimed at preventing bleeding in the initial phases by improving haemostatic functions and preventing thrombosis in the later resuscitation phases. Classic traditional interventions proven to target bleeding include DDVP/vasopressin, tranexamic acid, crystalloid minimization using plasma-based infusion, fibrinogen, and calcium. However, novel therapeutic targets are emerging that include nano-based semisynthetic platelets [148] but are still investigational.

12. Conclusion

Accidental or traumatic injuries are accompanied by changes in platelet function arising from endothelium and circulating factors alterations. Shortly after injury, platelets are dysfunctional characterized by increased expression of surface activation markers, decreased propensity to aggregate, and adherence to endothelial surfaces, which collectively increase the risk of bleeding. The hypofunction period that lasts for 72–96 hours is followed by restoration of function with attendant risk to thrombosis and vascular occlusion. The extent of platelet changes duration in each phase are dependent on modulating factors released during trauma and exogenously present either pre-trauma or added thereto. Unfortunately, *ex vivo* laboratory assays are neither predictive of functional performance nor designed to detect bleeding or thrombosis associated with trauma. A number of therapeutic interventions have been tried but lack clear equipoise.

Author details

Gordon Ogweno[1] and Edwin Murungi[2]*

1 Department of Medical Physiology, Kenyatta University, Kenya

2 Department of Medical Biochemistry, Kisii University, Kenya

*Address all correspondence to: edwinmurungi@kisiiuniversity.ac.ke

References

[1] Kottke-Marchant K, Corcoran G. The laboratory diagnosis of platelet disorders: An algorithmic approach. Archives of Pathology & Laboratory Medicine. 2002;**126**(2):133-146

[2] Fritsma GA. Platelet structure and function. Clinical Laboratory Science. 2015;**28**(2):125-131

[3] Brass L. Understanding and evaluating platelet function. Hematology ASH Education Program. 2010;**2010**(1):387-396. DOI: 10.1182/asheducation-2010.1.387

[4] Bye AP, Unsworth AJ, Gibbins JM. Platelet signaling: A complex interplay between inhibitory and activatory networks. Journal of Thrombosis and Haemostasis. 2016;**14**(5):918-930. DOI: 10.1111/jth.13302

[5] Estevez B, Du X. New concepts and mechanisms of platelet activation signaling. Physiology. 2017;**32**(2):162-177. DOI: 10.1152/physiol.00020.2016

[6] Heemskerk JWM, Bevers EM, Lindhout T. Platelet activation and blood coagulation. Thrombosis and Haemostasis. 2002;**88**:186-194

[7] Stalker TJ et al. Hierarchical organization in the hemostatic response and its relationship to the platelet-signaling network. Blood. 2013;**121**(10):1875-1885. DOI: 10.1182/blood-2012-09-457739

[8] Montague SJ, Andrews RK, Gardiner EE. Mechanisms of receptor shedding in platelets. Blood. 2018;**132**(24):2535-2545. DOI: 10.1182/blood-2018-03-742668

[9] Schriner JB et al. Platelet function in trauma: Is current technology in function testing missing the mark in injured patients? Shock. 2022;**58**(1):1-13. DOI: 10.1097/SHK.0000000000001948

[10] Nekludov M, Bellander BM, Blombäck M, Wallen HN. Platelet dysfunction in patients with severe traumatic brain injury. Journal of Neurotrauma. 2007;**24**(11):1699-1706. DOI: 10.1089/neu.2007.0322

[11] Brown LM, Call MS, Knudson MM, Cohen MJ, The Trauma Outcomes Group. A normal platelet count may not be enough: The impact of admission platelet count on mortality and transfusion in severely injured trauma patients. Journal of Trauma. 2011;**71** (2 SUPPL. 3):S337-S342. DOI: 10.1097/TA.0b013e318227f67c

[12] Stansbury LG, Hess AS, Thompson K, Kramer B, Scalea TM, Hess JR. The clinical significance of platelet counts in the first 24 hours after severe injury. Transfusion. 2013;**53**(4):783-789. DOI: 10.1111/j.1537-2995.2012.03828.x

[13] Jacoby RC, Owings JT, Holmes J, Battistella FD, Gosselin RC, Paglieroni TG. Platelet activation and function after trauma. The Journal of Trauma. 2001;**51**(4):639-647

[14] Vulliamy P et al. Loss of GPVI and GPIbα contributes to trauma-induced platelet dysfunction in severely injured patients. Blood Advances. 2020;**4**(12):2623-2630. DOI: 10.1182/bloodadvances.2020001776

[15] Le Blanc J, Mullier F, Vayne C, Lordkipanidzé M. Advances in platelet function testing—Light transmission aggregometry and beyond. Journal of Clinical Medicine. 2020;**9**(8):1-17. DOI: 10.3390/jcm9082636

[16] Ramsey MT et al. A prospective study of platelet function in trauma patients. Journal of Trauma and Acute Care Surgery. 2016;**80**(5):726-733. DOI: 10.1097/TA.0000000000001017

[17] St. AE, John et al, Platelets retain inducible alpha granule secretion by P-selectin expression but exhibit mechanical dysfunction during trauma-induced coagulopathy. Journal of Thrombosis and Haemostasis. 2019;**17**(5):771-781. DOI: 10.1111/jth.14414

[18] Roschitz B, Sudi K, Köstenberger M, Muntean W. Shorter PFA-100® closure times in neonates than in adults: Role of red cells, white cells, platelets and von Willebrand factor. Acta Paediatrica. 2001;**90**(6):664-670. DOI: 10.1111/j.1651-2227.2001.tb02431.x

[19] Dyszkiewicz-Korpanty AM, Frenkel EP, Sarode R. Approach to the assessment of platelet function: Comparison between optical-based platelet-rich plasma and impedance-based whole blood platelet aggregation methods. Clinical and Applied Thrombosis. 2005;**11**(1):25-35. DOI: 10.1177/107602960501100103

[20] Windeløv NA et al. Platelet aggregation following trauma: A prospective study. Blood Coagulation & Fibrinolysis. 2014;**25**:67-73. DOI: 10.1097/MBC.0b013e328364c2da

[21] Wade CE et al. Upon admission coagulation and platelet function in patients with thermal and electrical injuries. Burns. 2016;**42**(8):1704-1711. DOI: 10.1016/j.burns.2016.05.001

[22] Matthay ZA et al. Postinjury platelet aggregation and venous thromboembolism. Journal of Trauma and Acute Care Surgery. 2022;**93**(5):604-612. DOI: 10.1097/TA.0000000000003655

[23] Solomon C et al. Platelet function following trauma; a multiple electrode aggregometry study. Thrombosis and Haemostasis. 2011;**106**(2):322-330. DOI: 10.1160/TH11-03-0175

[24] Kutcher ME et al. Characterization of platelet dysfunction after trauma. Journal of Trauma and Acute Care Surgery. 2012;**73**(1):13-19. DOI: 10.1097/TA.0b013e318256deab

[25] McCully BH et al. Onset of coagulation function recovery is delayed in severely injured trauma patients with venous thromboembolism. Journal of the American College of Surgeons. 2017;**225**(1):42-51. DOI: 10.1016/j.jamcollsurg.2017.03.001

[26] Kornblith LZ et al. Perhaps it's not the platelet: Ristocetin uncovers the potential role of von Willebrand factor in impaired platelet aggregation following traumatic brain injury. Journal of Trauma and Acute Care Surgery. 2018;**85**(5):873-880. DOI: 10.1097/TA.0000000000002025

[27] Connelly CR et al. Assessment of three point-of-care platelet function assays in adult trauma patients. The Journal of Surgical Research. 2017;**212**:260-269. DOI: 10.1016/j.jss.2017.01.008

[28] Joseph B et al. A prospective evaluation of platelet function in patients on antiplatelet therapy with traumatic intracranial hemorrhage. Journal of Trauma and Acute Care Surgery. 2013;**75**(6):990-994. DOI: 10.1097/TA.0b013e3182a96591

[29] Glass NE, Riccardi J, Horng H, Kacprzynski G, Sifri Z. Platelet dysfunction in patients with traumatic intracranial hemorrhage: Do desmopressin and platelet therapy help or harm? American Journal of Surgery.

2022;**223**(1):131-136. DOI: 10.1016/j.amjsurg.2021.07.050

[30] Alvikas J et al. Rapid detection of platelet inhibition and dysfunction in traumatic brain injury: A prospective observational study. Journal of Trauma and Acute Care Surgery. 2021;**92**(1):167-176. DOI: 10.1097/TA.0000000000003427

[31] Sirajuddin S et al. Inhibition of platelet function is common following even minor injury. Journal of Trauma and Acute Care Surgery. 2016;**81**(2):328-332. DOI: 10.1097/TA.0000000000001057

[32] Wohlauer MV et al. Early platelet dysfunction: An unrecognized role in the acute coagulopathy of trauma. Journal of the American College of Surgeons. 2012;**214**(5):739-746. DOI: 10.1016/j.jamcollsurg.2012.01.050

[33] Kornblith LZ, Kutcher ME, Redick BJ, Calfee CS, Vilardi RF, Cohen MJ. Fibrinogen and platelet contributions to clot formation: Implications for trauma resuscitation and thromboprophylaxis. Journal of Trauma and Acute Care Surgery. 2014;**76**(2):255-263. DOI: 10.1097/TA.0000000000000108

[34] Mousa SA, Forsythe MS. Comparison of the effect of different platelet GPIIb/IIa antagonists on the dynamics of platelet/fibrin-mediated clot strength induced using Thromboelastography. Thrombosis Research. 2001;**104**:49-56

[35] Harr JN et al. Platelets are dominant contributors to hypercoagulability after injury. Journal of Trauma and Acute Care Surgery. 2013;**74**(3):756-765. DOI: 10.1097/TA.0b013e3182826d7e

[36] Verni CC, Davila A, Balian S, Sims CA, Diamond SL. Platelet dysfunction during trauma involves diverse signaling pathways and an inhibitory activity in patient-derived plasma. Journal of Trauma and Acute Care Surgery. 2019;**86**(2):250-259. DOI: 10.1097/TA.0000000000002140

[37] Sbrana S et al. Relationships between optical Aggregometry (type born) and flow cytometry in evaluating ADP-induced platelet activation. Cytometry Part B: Clinical Cytometry. 2008;**74B**:30-39. DOI: 10.1002/cyto.b.20360

[38] Matthay ZA et al. Dynamic effects of calcium on in vivo and ex vivo platelet behavior after trauma. Journal of Trauma and Acute Care Surgery. 2020;**89**(5):871-879. DOI: 10.1097/TA.0000000000002820

[39] Michelson AD et al. In vivo tracking of platelets: Circulating degranulated platelets rapidly lose surface P-selectin but continue to circulate and function. Proceedings of the National Academy of Sciences of the United States of America. 1996;**93**(21):11877-11882. DOI: 10.1073/pnas.93.21.11877

[40] Henn V et al. CD40 ligand on activated platelets triggers an inflammatory reaction of endothelial cells. Nature. 1998;**391**(6667):591-594. DOI: 10.1038/35393

[41] Chatterjee M, Gawaz M. Clinical significance of receptor shedding-platelet GPVI as an emerging diagnostic and therapeutic tool. Platelets. 2017;**28**(4):362-371. DOI: 10.1080/09537104.2016.1227062

[42] Kannan M, Ahmad F, Saxena R. Platelet activation markers in evaluation of thrombotic risk factors in various clinical settings. Blood Reviews. 2019;**37**:100583. DOI: 10.1016/j.blre.2019.05.007

[43] Moroi M, Farndale RW, Jung SM. Activation-induced changes in platelet

surface receptor expression and the contribution of the large-platelet subpopulation to activation. Research and Practice in Thrombosis and Haemostasis. 2020;**4**(2):285-297. DOI: 10.1002/rth2.12303

[44] Montague SJ et al. Soluble GPVI is elevated in injured patients: Shedding is mediated by fibrin activation of GPVI. Blood Advances. 2018;**2**(3):240-251. DOI: 10.1182/bloodadvances.2017011171

[45] Johansson PI et al. High sCD40L levels early after trauma are associated with enhanced shock, sympathoadrenal activation, tissue and endothelial damage, coagulopathy and mortality. Journal of Thrombosis and Haemostasis. 2012;**10**(2):207-216. DOI: 10.1111/j.1538-7836.2011.04589.x

[46] Caspers M et al. Microparticles profiling in trauma patients: High level of microparticles induce activation of platelets in vitro. European Journal of Trauma and Emergency Surgery. 2020;**46**(1):43-51. DOI: 10.1007/s00068-019-01111-7

[47] Fröhlich M et al. Temporal phenotyping of circulating microparticles after trauma: A prospective cohort study. Scandinavian Journal of Trauma, Resuscitation and Emergency Medicine. 2018;**26**(1, 1):-10. DOI: 10.1186/s13049-018-0499-9

[48] Windeløv NA et al. Low level of procoagulant platelet microparticles is associated with impaired coagulation and transfusion requirements in trauma patients. Journal of Trauma and Acute Care Surgery. 2014;**77**(5):692-700. DOI: 10.1097/TA.0000000000000437

[49] Dyer MR et al. Platelet-derived extracellular vesicles released after trauma promote hemostasis and contribute to DVT in mice. Journal of Thrombosis and Haemostasis. 2019;**17**(10):1733-1745. DOI: 10.1111/jth.14563

[50] Vulliamy P, Gillespie S, Armstrong PC, Allan HE, Warner TD, Brohi K. Histone H4 induces platelet ballooning and microparticle release during trauma hemorrhage. PNAS. 2019;**116**(35):17444-17449. DOI: 10.1073/pnas.1904978116

[51] Agbani EO et al. Coordinated membrane ballooning and procoagulant spreading in human platelets. Circulation. 2015;**132**(15):1414-1424. DOI: 10.1161/CIRCULATIONAHA.114.015036

[52] Agbani EO, Williams CM, Hers I, Poole AW. Membrane ballooning in aggregated platelets is synchronised and mediates a surge in Microvesiculation. Scientific Reports. 2017;**7**:2770. DOI: 10.1038/s41598-017-02933-4

[53] Wannberg M, Miao X, Li N, Wikman A, Wahlgren CM. Platelet consumption and hyperreactivity coexist in experimental traumatic hemorrhagic model. Platelets. 2020;**31**(6):777-783. DOI: 10.1080/09537104.2019.1678120

[54] Starr N, Matthay Z, Fields A, Neal MD, Kornblith LZ. Platelet transfusion for trauma resuscitation. Current Trauma Reports. 2022;**8**(3):147-159. DOI: 10.1007/s40719-022-00236-2

[55] Kornblith LZ, Moore HB, Cohen MJ. Trauma-induced coagulopathy: The past, present, and future. Journal of Thrombosis and Haemostasis. 2019;**17**(6):852-862. DOI: 10.1111/jth.14450

[56] Evans RJ, Gordon JL. Refractoriness in blood platelets: Effect of prior exposure to aggregating agents on subsequent aggregation responses.

British Journal of Pharmacology. 1974;**51**(1):123P

[57] Pareti FI, Capitanio A, Mannucci L, Ponticelli C, Mannucci PM. Acquired dysfunction due to the circulation of 'exhausted' platelets. The American Journal of Medicine. 1980;**69**(2):235-240. DOI: 10.1016/0002-9343(80)90383-6

[58] Fong BJSC, Kaplan BS. Impairment of platelet aggregation in hemolytic uremic syndrome: Evidence for platelet 'exhaustion'. Blood. 1982;**60**(3):564-571

[59] Dupuis A, Bordet JC, Eckly A, Gachet C. Platelet δ-storage pool disease: An update. Journal of Clinical Medicine. 2020;**9**(8):1-23. DOI: 10.3390/jcm9082508

[60] Ogweno GO. Importance of catecholamine signaling in the development of platelet exhaustion after traumatic injury comment. Journal of Thrombosis and Haemostasis. 2022:2715-2716. DOI: 10.1111/jth.15868

[61] Israels SJ, McNicol A, Robertson C, Gerrard JM. Platelet storage pool deficiency: Diagnosis in patients with prolonged bleeding times and normal platelet aggregation. British Journal of Haematology. 1990;**75**(1):118-121. DOI: 10.1111/j.1365-2141.1990.00118.x

[62] Gunning WT, Yoxtheimer L, Smith MR. Platelet aggregation assays do not reliably diagnose Platelet Delta granule storage Pool deficiency. Journal of Hematology. 2021;**10**(4):196-201. DOI: 10.14740/jh832

[63] Kornblith LZ, Robles AJ, Conroy AS, Miyazawa BY, Callcut RA, Cohen MJ. Tired platelet: Functional Anergy after injury. Journal of the American College of Surgeons. 2017;**225**(4):S64. DOI: 10.1016/j.jamcollsurg.2017.07.131

[64] Vulliamy P, Kornblith LZ, Kutcher ME, Cohen MJ, Brohi K, Neal MD. Alterations in platelet behavior after major trauma: Adaptive or maladaptive? Platelets. 2021;**32**(3):295-304. DOI: 10.1080/09537104.2020.1718633

[65] Hosseini E, Mohtashami M, Ghasemzadeh M. Down-regulation of platelet adhesion receptors is a controlling mechanism of thrombosis, while also affecting post-transfusion efficacy of stored platelets. Thrombosis Journal. 2019;**17**(1):1-11. DOI: 10.1186/s12959-019-0209-5

[66] Johansson PI et al. Traumatic endotheliopathy: A prospective observational study of 424 severely injured patients. Annals of Surgery. 2017;**265**(3):597-603. DOI: 10.1097/SLA.0000000000001751

[67] Britten MW, Lümers L, Tominaga K, Peters J, Dirkmann D. Glycocalyx components affect platelet function , whole blood coagulation, and fibrinolysis: An in vitro study suggesting a link to trauma-induced coagulopathy. BMC Anesthesiology. 2021;**21**:83

[68] Starr NE et al. Identification of injury and shock driven effects on ex vivo platelet aggregometry: A cautionary tale of phenotyping. Journal of Trauma and Acute Care Surgery. 2020;**89**(1):20-28. DOI: 10.1097/TA.0000000000002707

[69] South K, Freitas MO, Lane DA. Conformational quiescence of ADAMTS-13 prevents proteolytic promiscuity. Journal of Thrombosis and Haemostasis. 2016;**14**(10):2011-2022. DOI: 10.1111/jth.13445

[70] Clark CC et al. Truncation of ADAMTS13 by plasmin enhances its activity in plasma. Thrombosis and Haemostasis. 2018;**118**(3):471-479. DOI: 10.1055/s-0038-1627460

[71] Rourke C et al. Fibrinogen levels during trauma hemorrhage, response to replacement therapy, and association with patient outcomes. Journal of Thrombosis and Haemostasis. 2012;**10**(7):1342-1351. DOI: 10.1111/j.1538-7836.2012.04752.x

[72] Macarthur TA et al. Quantification of von Willebrand factor and ADAMTS-13 after traumatic injury: A pilot study. Trauma Surgery & Acute Care Open. 2021;**6**(1):1-7. DOI: 10.1136/tsaco-2021-000703

[73] Russell RT, Mcdaniel JK, Pittet XLZJ, Shroyer M, Wagener BM. Low plasma ADAMTS13 activity is associated with coagulopathy, endothelial cell damage and mortality after severe Paediatric trauma. Thrombosis and Haemostasis. 2018;**118**(4):676-687

[74] Dyer MR et al. Traumatic injury results in prolonged circulation of ultralarge von Willebrand factor and a reduction in ADAMTS13 activity. Transfusion. 2020;**60**(6):1308-1318. DOI: 10.1111/trf.15856

[75] Siemiatkowski A, Kloczko J, Galar M, Czaban S. von Willebrand factor antigen as a prognostic marker in posttraumatic acute lung injury. Hemostasis. 2000;**30**:189-195

[76] Michelson AD. Thrombin-induced Down-regulation of the platelet membrane glycoprotein Ib-IX complex. Seminars in Thrombosis and Hemostasis. 1992;**18**(1):23-25

[77] Dunbar NM, Chandler WL. Thrombin generation in trauma patients. Transfusion. 2009;**49**(12):2652-2660. DOI: 10.1111/j.1537-2995.2009.02335.x

[78] Yanagida Y et al. Normal prothrombinase activity, increasedsystemic thrombin activity, and lower antithrombin levels in patients with disseminated intravascular coagulation at an early phase of trauma: Comparison with acute coagulopathy of trauma-shock. Surgery (United States). 2013;**154**(1):48-57. DOI: 10.1016/j.surg.2013.02.004

[79] Goswami J et al. Neutrophil extracellular trap formation and Syndecan-1 shedding are increased after trauma. Shock. 2021;**56**(3):433-439. DOI: 10.1097/SHK.0000000000001741

[80] Kornblith LZ et al. It's about time: Transfusion effects on postinjury platelet aggregation over time. Journal of Trauma and Acute Care Surgery. 2019;**87**(5):1042-1051. DOI: 10.1097/TA.0000000000002459

[81] Moore HB, Moore EE. Temporal changes in fibrinolysis following injury. Seminars in Thrombosis and Hemostasis. 2020;**46**(2):189-198. DOI: 10.1055/s-0039-1701016

[82] Napolitano F. Role of plasminogen activation system in platelet pathophysiology: Emerging concepts for translational applications. International Journal of Medical Sciences. 2022;**23**:6065. DOI: 10.3390/ijms23116065

[83] Blockmans D et al. The effect of plasmin on platelet function the effect of plasmin on platelet function. Platelets. 2015;7:139-148. DOI: 10.3109/09537109609023572

[84] Verni CC, Davila A, Sims CA, Diamond SL. D-dimer and fibrin degradation products impair platelet signaling: Plasma D-dimer is a predictor and mediator of platelet dysfunction during trauma. The Journal of Applied Laboratory Medicine. 2020;**5**(6):1253-1264. DOI: 10.1093/jalm/jfaa047

[85] Schafer AI, Zavoico GB, Loscalzo J, Maas AK. Synergistic inhibition of platelet activation by plasmin and prostaglandin I2. Blood. 1987;**69**(5):1504-1507. DOI: 10.1182/blood.V69.5.1504.1504

[86] Lu J, Hu P, Wei G, Luo Q, Qiao J, Geng D. Effect of alteplase on platelet function and receptor expression. The Journal of International Medical Research. 2019;**47**(4):1731-1739. DOI: 10.1177/0300060519829991

[87] Adelman BB, Rizk A, Hanners E. Plasminogen interactions with platelets in plasma. Blood. 1988;**72**(5):1530-1535. DOI: 10.1182/blood.V72.5.1530.1530

[88] Whyte CS et al. Plasminogen associates with phosphatidylserine-exposing platelets and contributes to thrombus lysis under flow. Blood. 2015;**125**(16):23-26. DOI: 10.1182/blood-2014-09-599480

[89] Gouin I, Lecompte T, Morel M, Lebrazi J. In vitro effect of plasmin on human platelet function in plasma inhibition of aggregation caused by Fibrinogenolysis. Circulation. 1992;**85**:935-941. DOI: 10.1161/01.CIR.85.3.935

[90] Van Der Vorm LN, Remijn JA, De Laat B. Effects of plasmin on von Willebrand factor and platelets: A narrative review. TH Open. 2018;**2**(2):218-228

[91] Crawley JTB, Lam JK, Rance JB, Mollica LR, O'Donnell JS, Lane DA. Proteolytic inactivation of ADAMTS13 by thrombin and plasmin. Blood. 2005;**105**(3):1085-1093. DOI: 10.1182/blood-2004-03-1101

[92] Pielsticker C, Brodde MF, Raum L, Jurk K, Kehrel BE. Plasmin-induced activation of human platelets is modulated by Thrombospondin-1, Bona fide misfolded proteins and thiol isomerases. International Journal of Molecular Sciences. 2020;**21**(22):8851. DOI: 10.3390/ijms21228851

[93] Fletcher-Sandersjöö A, Thelin EP, Maegele M, Svensson M, Bellander BM. Time course of hemostatic disruptions after traumatic brain injury: A systematic review of the literature. Neurocritical Care. 2021;**34**(2):635-656. DOI: 10.1007/s12028-020-01037-8

[94] Roberts DJ et al. Time course and outcomes associated with transient versus persistent fibrinolytic phenotypes after injury: A nested, prospective, multicenter cohort study. Journal of Trauma and Acute Care Surgery. 2019;**86**(2):206-213. DOI: 10.1097/TA.0000000000002099

[95] Brogren H, Wallmark K, Denium J, Karlsson L, Jern S. Platelets retain high levels of active plasminogen activator inhibitor 1. PLos One. 2011;**6**(11):e26762. DOI: 10.1371/journal.pone.0026762

[96] Timmermans K et al. Plasma levels of danger-associated molecular patterns are associated with immune suppression in trauma patients. Intensive Care Medicine. 2016;**42**(4):551-561. DOI: 10.1007/s00134-015-4205-3

[97] Vourc'h M, Roquilly A, Asehnoune K. Trauma-induced damage-associated molecular patterns-mediated remote organ injury and immunosuppression in the acutely ill patient. Frontiers in Immunology. 2018;**9**(Jun). DOI: 10.3389/fimmu.2018.01330

[98] Ottestad W et al. HMGB1 concentration measurements in trauma patients: Assessment of pre-analytical conditions and sample material. Molecular Medicine. 2019;**26**(1):1-8. DOI: 10.1186/s10020-019-0131-0

[99] Relja B, Land WG. Damage-associated molecular patterns in trauma.

European Journal of Trauma and Emergency Surgery. 2020;**46**(4):751-775. DOI: 10.1007/s00068-019-01235-w

[100] Etulain J, Martinod K, Wong SL, Cifuni SM, Schattner M, Wagner DD. P-selectin promotes neutrophil extracellular trap formation in mice. Blood. 2015;**126**(2):242-246. DOI: 10.1182/blood-2015-01-624023

[101] Semeraro F et al. Extracellular histones promote thrombin generation through platelet-dependent mechanisms: Involvement of platelet TLR2 and TLR4. Blood. 2011;**118**(7):1952-1961. DOI: 10.1182/blood-2011-03-343061

[102] Fields AT et al. Good platelets gone bad: The effects of trauma patient plasma on healthy platelet aggregation. Shock. 2021;**55**(2):189-197. DOI: 10.1097/SHK.0000000000001622

[103] Wiener G et al. Shock releases bile acid-inducing platelet inhibition and fibrinolysis. The Journal of Surgical Research. 2015;**no. February**:1-6. DOI: 10.1016/j.jss.2015.01.046

[104] Lawson CA, Spangler EA. The effect of adding lactic acid to canine whole blood on platelet aggregation as measured by impedance aggregometry. Veterinary Clinical Pathology. 2020;**49**(2):217-221. DOI: 10.1111/vcp.12862

[105] Green FW, Kaplan MM, Curtis LE, Levine PH. Effect of acid and pepsin on blood coagulation and platelet aggregation. A possible contributor to prolonged gastroduodenal mucosal hemorrhage. Gastroenterology. 1978;**74**(1):38-43. DOI: 10.1016/0016-5085(78)90352-9

[106] Rogers AB. The effect of pH on human platelet aggregation induced by epinephrine and ADP (36307). Experimental Biology and Medicine. 1972;**139**(4):1100-1103

[107] Marumo M, Suehiro A, Kakishita E, Groschner K, Wakabayashi I. Extracellular pH affects platelet aggregation associated with modulation of store-operated Ca2+ entry. Thrombosis Research. 2001;**104**(5):353-360. DOI: 10.1016/S0049-3848(01)00374-7

[108] Etulain J et al. Acidosis downregulates platelet haemostatic functions and promotes neutrophil proinflammatory responses mediated by platelets. Thrombo Haemost. 2012;**107**:99-110. DOI: 10.1160/TH11-06-0443

[109] Binkowska AM, Michalak G, Pilip S, Kopacz M, Słotwiński R. The diagnostic value of early cytokine response in patients after major trauma – Preliminary report. Clinical Immunology. 2018;**43**(1):33-41

[110] Wallen TE et al. Survival analysis by in fl ammatory biomarkers in severely injured patients undergoing damage control resuscitation. Surgery. 2022;**171**(3):818-824. DOI: 10.1016/j.surg.2021.08.060

[111] Brøchner AC, Toft P. Pathophysiology of the systemic inflammatory response after major accidental trauma. Scandinavian Journal of Trauma, Resuscitation and Emergency Medicine. 2009;**17**(1):1-10. DOI: 10.1186/1757-7241-17-43

[112] Oleksowicz L, Owiec Z, Isaacs R, Dutcher JP, Puszkin E. Morphologic and ultrastructural evidence for interleukin-6 induced platelet activation. American Journal of Hematology. 1995;**48**(2):92-99. DOI: 10.1002/ajh.2830480205

[113] Lumadue JA, Lanzkron SM, Kennedy SD, Kuhl DT, Mt BS, Kickler TS.

Cytokine induction of platelet activation. American Journal of Clinical Pathology. 1996;**106**:795-798

[114] Burstein SA et al. Cytokine-induced alteration of platelet and hemostatic function. Stem Cells. 1996;**14**(suppl 1):154-162

[115] Atefi G et al. Complement activation in trauma patients alters platelet function. Shock. 2016;**46**(3S):83-88. DOI: 10.1097/SHK.0000000000000675

[116] Sillesen M et al. Assessment of coagulopathy, endothelial injury, and inflammation after traumatic brain injury and hemorrhage in a porcine model. Journal of Trauma and Acute Care Surgery. 2014;**76**(1):12-20. DOI: 10.1097/TA.0b013e3182aaa675

[117] Anfossi G, Trovati M. Role of catecholamines in platelet function: Pathophysiological and clinical significance. European Journal of Clinical Investigation. 1996;**26**(5):353-370. DOI: 10.1046/j.1365-2362.1996.150293.x

[118] Di Battista AP et al. Sympathoadrenal activation is associated with acute traumatic coagulopathy and Endotheliopathy in isolated brain injury. Shock. 2016;**46**(3S):96-103. DOI: 10.1097/SHK.0000000000000642

[119] Davis JW, Phillips PE. The effect of ethanol on human platelet aggregation in vitro. Atherosclerosis. 1970;**11**:473-477

[120] Haut MJ, Cowan DH. The effect of ethanol on hemostatic properties of human blood platelets. The American Journal of Medicine. 1974;**56**:22-57

[121] Renaud SC, Ruf J. Effects of alcohol on platelet functions. Clinica Chimica Acta. 1996;**246**:77-89

[122] Elmer O, Goransson G, Zoucas E. Impairment of primary hemostasis and platelet function after alcohol ingestion in man. Hemostasis. 1984;**14**:223-228

[123] Schramko A, Niemi T. Alcohol induces profuse reversible decrease in platelet function and delayed initiation of co-agulation. Emergency Medicine Health Care. 2014;**2**:3-5. DOI: 10.7243/2052-6229-2-4

[124] Elmer O, Zoucas E, Bengmark S. Effects of ethanol on platelet aggregation: An in vitro study *. Res Exp Med(Berl). 1983;**182**(98):13-19

[125] Wang I-J et al. Effect of acute alcohol intoxication on mortality, coagulation, and fibrinolysis in trauma patients. PLoS One. 2021;**16**:e0248810. DOI: 10.1371/journal.pone.02

[126] Howard BM et al. Exposing the bidirectional effects of alcohol on coagulation in trauma: Impaired clot formation and decreased fibrinolysis in rotational thromboelastometry. Journal of Trauma and Acute Care Surgery. 2018;**84**(1):97-103. DOI: 10.1097/TA.0000000000001716

[127] Pieters M, Vorster HH, Jerling JC, Venter CS, Kotze RCM, Bornman E. The effect of ethanol and its metabolism on fibrinolysis. Thrombosis and Haemostasis. 2010;**104**(9):724-733. DOI: 10.1160/TH10-01-0048

[128] Wallen TE et al. Platelet dysfunction persists after trauma despite balanced blood product resuscitation. Surgery. 2022;**xxx**:1-9. DOI: 10.1016/j.surg.2022.09.017

[129] Goodman JH. Platelet aggregation in experimental spinal cord injury. Archives of Neurology. 1979;**36**(4):197. DOI: 10.1001/archneur.1979.00500400051006

[130] Matsumoto H et al. ADAMTS13 activity decreases in the early phase of

trauma associated with coagulopathy and systemic inflammation: A prospective observational study. Thrombosis Journal. 2021;**19**(1):1-8. DOI: 10.1186/s12959-021-00270-1

[131] Vecht CJ, Minderhoud JM, Sibinga CTS. Platelet Aggregability in relation to impaired consciousness after head injury. Journal of Clinical Pathology. 1975;**28**:814-820

[132] Ersoz G, Ficicilar H, Pasin M, Yorgancioglu R, Yavuzer S. Platelet aggregation in traumatic spinal cord injury. Spinal Cord. 1999;**37**(9):644-647. DOI: 10.1038/sj.sc.3100903

[133] Martin GE et al. Platelet function changes in a time-dependent manner following traumatic brain injury in a murine model. Shock. 2018;**50**(5):551-556. DOI: 10.1097/SHK.0000000000001056

[134] Sumislawski JJ, Kornblith LZ, Conroy AS, Callcut RA, Cohen MJ. Dynamic coagulability after injury: Is delaying venous thromboembolism chemoprophylaxis worth the wait? Journal of Trauma and Acute Care Surgery. 2018;**85**(5):907-914. DOI: 10.1097/TA.0000000000002048

[135] Hamada SR et al. Impact of platelet transfusion on outcomes in trauma patients. Critical Care. 2022;**26**(1):49. DOI: 10.1186/s13054-022-03928-y

[136] Schnuriger B et al. The impact of platelets on the progression of traumatic. The Journal of Trauma. 2010;**68**(4):881-885. DOI: 10.1097/TA.0b013e3181d3cc58

[137] Geerts WH, Code KI, Jay RM, Chen E, Szalai JP. A Propspective study of venous thromboembolism after major trauma. The New England Journal of Medicine. 1994;**331**(24):1601-1606

[138] Nielsen S et al. Early detection of deep venous thrombosis in trauma patients. Cureus. 2020;**12**(7):1-9. DOI: 10.7759/cureus.9370

[139] Rossi EC, Green D, Rosen JS, Spies SM, Jao JST. Sequential changes in factor VIII and platelets preceding deep vein thrombosis in patients with spinal cord injury. British Journal of Haematology. 1980;**45**(1):143-151. DOI: 10.1111/j.1365-2141.1980.tb03819.x

[140] Saillant NN, Sims CA. Platelet dysfunction in injured patients. Molecular Cell. Therapy. 2014;**2**(1):1-7. DOI: 10.1186/s40591-014-0037-8

[141] Guillotte AR et al. Effects of platelet dysfunction and platelet transfusion on outcomes in traumatic brain injury patients brain injury patients. Brain Injury. 2018;**52**(13-14):1849-1857. DOI: 10.1080/02699052.2018.1536805

[142] Henriksen HH et al. Impact of blood products on platelet function in patients with traumatic injuries: A translation study. The Journal of Surgical Research. 2017;**214**(March):154-161. DOI: 10.1016/j.jss.2017.02.037

[143] Cardenas JC et al. Platelet transfusions improve hemostasis and survival in a substudy of the prospective, randomized PROPPR trial. Blood Advances. 2018;**2**(14):1696-1704. DOI: 10.1182/bloodadvances.2018017699

[144] Vulliamy P, Gillespie S, Gall LS, Green L, Brohi K, Davenport RA. Platelet transfusions reduce fibrinolysis but do not restore platelet function during trauma hemorrhage. Journal of Trauma and Acute Care Surgery. 2017;**83**(3):388-397. DOI: 10.1097/TA.0000000000001520

[145] Briggs A, Gates JD, Kaufman RM, Calahan C, Gormley WB, Havens JM.

Platelet dysfunction and platelet transfusion in traumatic brain injury. The Journal of Surgical Research. 2015;**193**(2):802-806. DOI: 10.1016/j.jss.2014.08.016

[146] Ji H, Lee Y, Ha Y, Kim K, Koo K. Little impact of antiplatelet agents on venous thromboembolism after hip fracture surgery. Journal of Korean Medicine. 2011;**26**:1625-1629

[147] Brill JB et al. Aspirin as added prophylaxis for deep vein thrombosis in trauma. Journal of Trauma Acute Care Surgery. 2016;**80**(4):625-630. DOI: 10.1097/TA.0000000000000977

[148] Sloos PH et al. Platelet dysfunction after trauma: From mechanisms to targeted treatment. Transfusion. 2022;**62**(S1):S281-S300. DOI: 10.1111/trf.16971

Chapter 3

Lower Extremity Artery Disease (LEAD)

Yuki Matsumoto

Abstract

Recently, there has been significant progress in finalizing devices for lower extremity artery disease. Especially in the region of the superficial femoral artery, it is possible to benefit from drug technology. It is necessary to select a device that is appropriate for the lesion, lesion length, and patient background. On the other hand, there are still issues in the treatment of chronic limb ischemia and below-the-knee arteries. In the first place, the pathologies of "above the knee" and "below the knee" are different, and the purpose of treatment is also different. Access sites for treatment have also become smaller and more diverse with the development of peripheral devices.

Keywords: LEAD, EVT, DES, covered stent, DCB, BTK, BTA

1. Introduction

More than any other class of human disease, diseases of the arterial system are responsible for morbidity and mortality. Malformations of the vasculature are the underlying cause of clinical disease, which may present as either a narrowing or complete obstruction of the vessel lumen or as a progressive or sudden worsening.

Peripheral arterial disease is classified into carotid, vertebral, subclavian, renal, iliac, upper and lower extremity arteries, and treatment methods have been established for each region in recent years. Previously, peripheral arterial disease was referred to as arteriosclerosis obliterans, but since symptoms can appear without obstruction and can be severe, the term "lower extremity arterial disease (LEAD)" has come to be used generically. The number of LEAD is increasing due to the progression of atherosclerotic lesions. The increase in LEAD has led to the development of a wide range of treatment methods and the emergence of a variety of new treatment devices. Advances in technology have significantly improved patency rates, benefiting many patients.

2. Epidemiology and risk factors

2.1 Epidemiology

LEAD is an atherosclerotic disease that affects approximately 200 million people worldwide, with symptoms beginning at age 50 or older, and the incidence increasing rapidly after age 65. The incidence of LEAD increases dramatically after the age of 65, and the incidence rate exceeds 20% in people over the age of 80 [1]. LEAD tends to be higher in high-income populations and lower in low- to middle-income populations due to arteriosclerotic disease. However, a gender-specific analysis shows that LEAD is more prevalent among men in high-income populations and among women in low- to middle-income populations. There is now a growing consensus that the most reliable source of epidemiological data for assessing LEAD prevalence is ankle-brachial index (ABI) measurements (**Figure 1**). In a German study, about 18% of the patients were below 0.9, but only about 10% of them actually had claudication symptoms, and other reports have put the rate at 20–30% [2–4]. There were also racial differences, with blacks having higher incidence rates than non-Hispanic whites and blacks, and among all races, blacks of both male and female had higher rates of ABI < 0.9. Hispanic, Indian, and non-Hispanic whites were all comparable.

2.2 Risk factor

The main cause of LEAD is atherosclerosis. Risk factors for LEAD include all factors that contribute to the development of atherosclerosis.

2.2.1 Age

Susceptibility to developing atherosclerosis increases with age. Symptoms typically appear after the age of 40 [5].

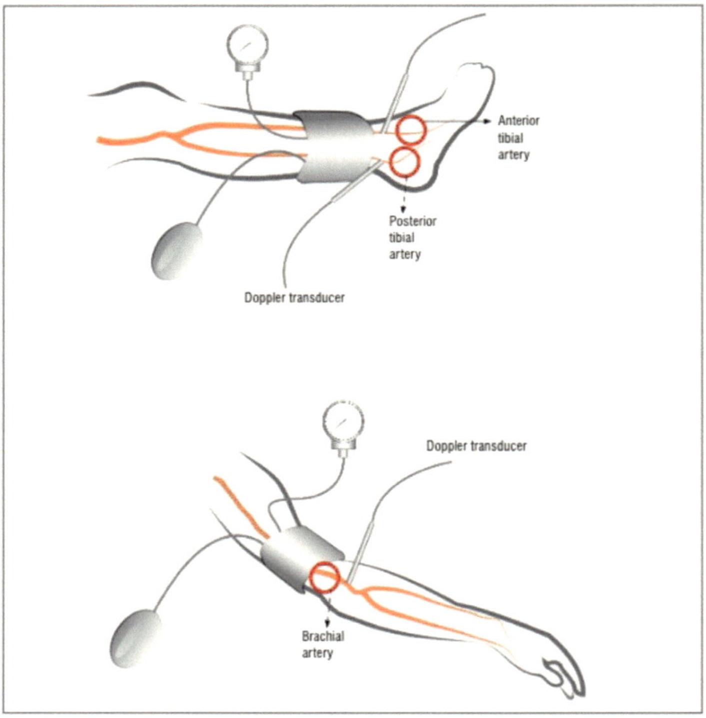

Figure 1.
Principles of correct measurement of the ankle brachial index.

DOI: http://dx.doi.org/10.5772/intechopen.1001354

2.2.2 Gender

All other factors being equal, men are much more likely than women to be affected by atherosclerosis and its consequences. Myocardial infarction and other complications of atherosclerosis are uncommon in premenopausal women, unless they have a predisposition through diabetes, hyperlipidemia, or severe hypertension. After the menopause, however, the incidence of atherosclerosis-related diseases increases due to hormone changes. This is probably due to a decrease in natural estrogen levels. The incidence of myocardial infarction becomes the same for men and women by the seventh to eighth decade of life [5].

2.2.3 Genetics

There is a well-established familial predisposition to atherosclerosis and ischemic heart disease, which is most likely polygenic in nature. Most commonly, the genetic predisposition relates to a familial clustering of other risk factors, such as high blood pressure or diabetes, while less commonly it relates to a well-defined inherited genetic abnormality in lipoprotein metabolism that results in excessively elevated blood lipid levels, such as familial hypercholesterolemia.

2.2.4 Hyperlipidemia

Hyperlipidemia is an important risk factor for atherosclerosis. Hypercholesterolemia is the focus of most of the evidence. Even in the absence of other risk factors, elevated serum cholesterol alone is sufficient to promote lesion development. Low-density lipoprotein (LDL) cholesterol, which plays an important physiological role as a vehicle for transporting cholesterol to peripheral tissues, is the component of total serum cholesterol associated with increased risk. HDL, on the other hand, is thought to mobilize cholesterol from developing and existing arteriosclerosis and transport it to the liver to be excreted into the bile. High HDL levels are associated with a lower risk. As a result, there is a great deal of interest in dietary, pharmacological and behavioral methods to reduce LDL levels and increase HDL levels. Regular exercise and moderate alcohol consumption both increase HDL levels, while obesity and smoking decrease it. Statins lower circulating cholesterol indirectly by inhibiting HMG-CoA reductase, a key enzyme in cholesterol biosynthesis in the liver [6–8].

2.2.5 Hypertension

High blood pressure is a significant risk factor for atherosclerosis at all ages. Men aged 45–62 years with blood pressure above 169/95 mm Hg have more than five times the risk of ischemic heart disease as those with blood pressure 140/90 mm Hg or lower. Both systolic and diastolic levels increase risk. Antihypertensive therapy reduces the incidence of diseases related to atherosclerosis, in particular stroke and ischemic heart disease. High blood pressure was the strongest indicator of all acute forms of PAD, including acute limb ischemia, chronic limb-threatening ischemia (CLTI), and their respective treatment outcomes [9].

2.2.6 Cigarette smoking

A particularly strong risk factor for LEAD [9]. Its contribution to the total risk in the population has been estimated to be around 44%. Notably, the association between LEAD and smoking persists even after smoking cessation, although it is significantly reduced more than 10 years after quitting [10].

2.2.7 Diabetes mellitus

Similar to smoking, diabetes is a strong risk factor for LEAD. The longer a person has diabetes, the higher the rate of LEAD. It also increases the risk of amputation of the lower extremity, as well as the risk of wound infection. Wound infections after amputation are also a risk for sepsis and often require additional amputations [11].

2.2.8 Other risk factors

Inflammation plays an important pathophysiological role in atherosclerotic disease. Several markers of inflammation, including high-sensitivity C-reactive protein, fibrinogen, and interleukin-6, are associated with increased risk of LEAD incidence, progression, and complications. Some autoimmune and inflammatory disorders, such as systemic lupus erythematosus and rheumatoid arthritis, also confer an increased risk of LEAD. Homocysteine, a non-protein amino acid, has also been shown to provide weak additive prognostic information in addition to the standard lipid measures. Several genetic factors have been identified as potential risk factors for atherosclerosis, though the evidence for their clinical significance remains weak [12].

3. Diagnostic approach

A comprehensive personal and family medical history should be taken and reviewed. Family history of coronary artery disease (CAD), cerebrovascular disease, aortic aneurysm, and LEAD should be reviewed. A thorough assessment of cardiovascular risk factors and co-morbidities, as well as a review of symptoms in relation to different vascular territories, should also be part of the clinical history. Systematic assessment of lifestyle, diet, walking, and physical activity is needed. Levels of physical activity should be assessed, and questionnaires and measures of functional status can provide reasonably accurate outcome measures. These can be useful in determining the level of impairment and in selecting appropriate care.

4. Diagnostic methods for LEAD

4.1 Ankle-brachial index (ABI)

ABI is a non-invasive tool that is useful in diagnosing and monitoring PAD. It is also a strong marker of generalized atherosclerosis and of the risk of cardiovascular disease. An ABI of 1.40 represents atherosclerosis and is associated with increased risk of cardiovascular events and mortality. It is more common in older patients and in particular in patients with diabetes or those with chronic kidney disease (CKD). When added to a risk score, the ABI can be used to improve the estimate of risk in one third and one fifth of women and men at "low risk", respectively. It is a valid method for the assessment of cardiovascular risk in a variety of ethnic groups, independent of the risk factors. In contrast to the coronary calcium score and the measurement of the intima-media thickness of the carotid arteries, the ABI is inexpensive and relatively quick to perform. It is important that the test is performed by trained staff. In addition to assessing general cardiovascular risk, ABI measurement can identify a patient's risk for events in the lower extremities, which requires close attention and education for the prevention of foot wounds [13].

4.2 Duplex ultrasounds (DUS)

For both screening and diagnostic purposes, duplex ultrasound (DUS) is often the first step in the vascular examination. To detect and localize vascular lesions and quantify their extent and severity using velocity criteria, DUS includes B-mode echography, pulsed wave, continuous wave, color, and power Doppler modalities (**Figure 2**). More recent techniques, such as flow imaging or live three-dimensional (3D) echography, as well as the use of ultrasonic contrast agents, are further improving the capabilities of DUS, although their use is still limited. DUS can be used to detect subclinical arterial disease, which is important for the assessment of CV risk. DUS is also effective for endovascular treatment because it can draw images in real time. However, the skill of the examiner greatly affects image rendition, making training important.

4.3 Computed tomography angiography (CTA)

Multidetector computed tomography angiography (MDCTA) boasts a succinct examination duration while minimizing motion and respiration artifacts during the imaging of vessels and organs (**Figure 3**). The advantages of CT angiography (CTA) include its non-invasive acquisition, wide availability, high resolution, and ability to reformat images in three dimensions. Similar to digital subtraction angiography (DSA) and magnetic resonance angiography (MRA), CTA provides a complete "roadmap" of the vasculature, which is critical for determining interventional strategies, including the location and seriousness of the lesion, as well as the upstream and downstream status [14]. However, limitations of CTA include lack of functional and

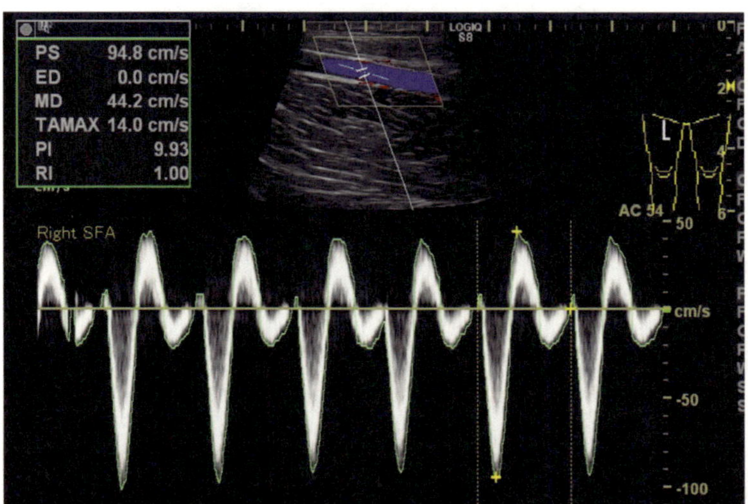

Figure 2.
Imaging of duplex ultrasound–courtesy of author.

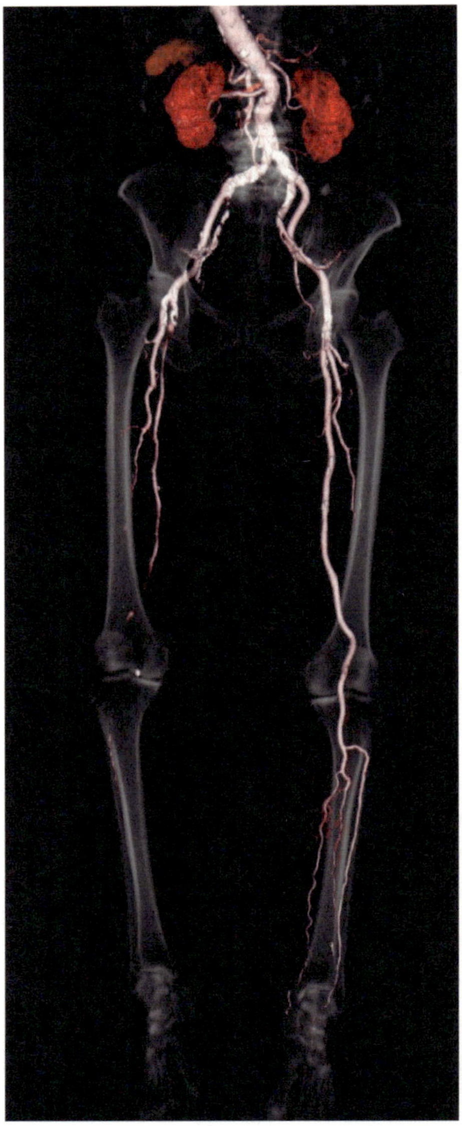

Figure 3.
Computed tomography angiography scan–courtesy of author.

hemodynamic data, radiation exposure, and use of iodinated contrast, which should be limited in patients with chronic kidney disease (CKD) and used cautiously in allergic patients. Nephrotoxicity can be reduced by minimizing contrast agent volume and ensuring adequate hydration prior to and following imaging. The efficacy of acetylcysteine in limiting nephrotoxicity is uncertain. Recent studies have proposed that statins or sodium bicarbonate may prevent contrast agent-induced nephrotoxicity.

4.4 Magnetic resonance angiography (MRA)

Magnetic resonance angiography (MRA) is a non-invasive, operator-independent imaging modality akin to CTA. High-fidelity three-dimensional (3D) reconstructions may be generated with high sensitivity and specificity of 86% and 93%, respectively, in comparison with conventional angiography. An advantage of this technique is the utilization of non-ionic contrast agents, such as gadolinium, which are considerably less nephrotoxic and 6–8 times less allergenic (1%). However, notable limitations of MRA include the prolonged examination duration, the potential for claustrophobia, contraindications in patients with pacemakers or other metallic implants, and the significant impact of motion artifacts. MRA scans may also be acquired without the administration of contrast agents, but this is associated with a decreased visualization of lesions and an increased number of artifacts (**Figure 4**).

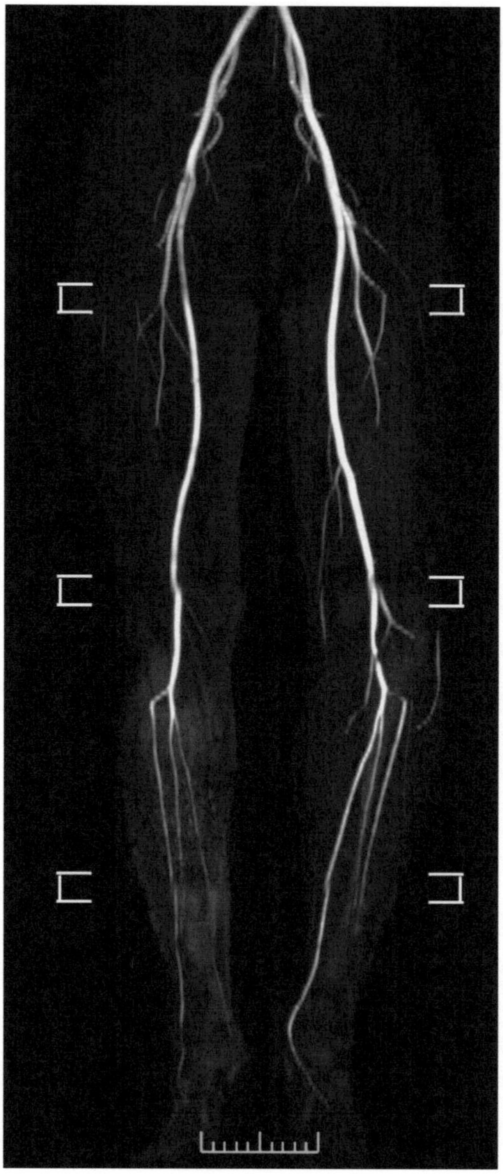

Figure 4.
Magnetic resonance angiography scan–courtesy of author.

4.5 Digital subtraction angiography (DSA)

Digital subtraction angiography (DSA) has long been considered the gold standard in vascular imaging (**Figure 5**). But its use has decreased because of its invasive nature and the possibility of serious, possibly lethal, complications. The incidence of iatrogenic complications such as hematoma, pseudoaneurysm, arteriovenous fistula,

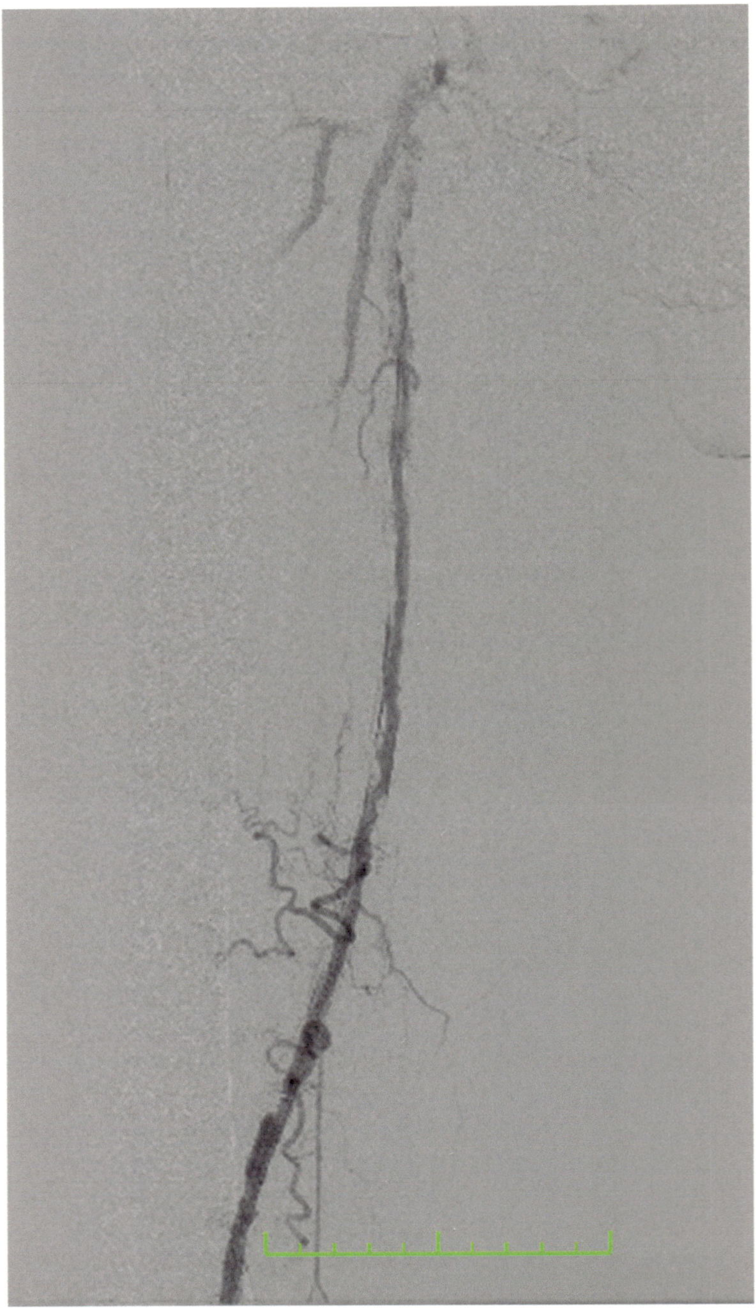

Figure 5.
Digital subtraction angiography–courtesy of author.

arterial thrombosis, and contrast media-related complications has been reported to be 0.7% with a mortality rate of 0.16% [5]. Despite these disadvantages, DSA remains a preferred imaging technique because it provides excellent resolution, particularly in small caliber arteries. Furthermore, it allows visualization of the collateral circulation and measurement of pressure gradients to assess the hemodynamic significance of the stenoses being studied.

The advantage of subtractive angiography is that it combines the diagnostic evaluation with the therapeutic intervention. In patients with intermittent claudication, it should only be performed as part of a single interventional procedure.

5. Treatment approach

5.1 Smoking cessation

A body of research attests to the advantages of smoking cessation in reducing cardiovascular events and mortality, particularly in patients with cerebrovascular disease and peripheral arterial disease [15]. The 2016 European Society of Cardiology (ESC) guidelines on cardiovascular disease prevention extensively address the management and support of smoking cessation. Additionally, exposure to secondhand smoke should be evaluated and mitigated.

5.2 Lipid-lowering drugs

The same guidelines for lipid-lowering treatment apply to all individuals with PAD, including those undergoing medical therapy, those preparing for surgical or endovascular procedures, and those undergoing secondary prevention. Lipid-lowering treatment is essential for all patients with PAD because it extends life, reduces the risk of cardiovascular and cerebrovascular events, and decreases the need for endovascular revascularization and the incidence of amputation. Lipid profile analysis is recommended for every patient with PAD and should be repeated at least annually to assess achievement of LDL-C targets. Assessments at one-two months intervals are recommended if treatment needs to be modified and/or target levels are not achieved. LDL-C targets should be considered when evaluating the results of lipid-lowering treatment; once achieved, secondary therapeutic goals may be pursued by adding appropriate omega-3 fatty acids and/or fibrates to non-pharmacological management.

5.3 Antithrombotic therapy in lower extremity artery disease

Antiplatelet therapy is indicated in patients with any atherosclerotic lesion, including LEAD, where intermittent claudication is an indication for antiplatelet therapy. Both surgical and endovascular therapy require pre- and postoperative antiplatelet or anticoagulant therapy. Similar to the prevention of ischemic heart disease, ischemic events in the lower extremities are treated with similar pharmacologic therapies. Although guidelines describe the management of antiplatelet or anticoagulant therapy, a recent Voyager PAD publication demonstrated that anticoagulant therapy has no inferiority over aspirin alone. Depending on the patient's condition, either single or dual antiplatelet therapy should be selected, and the patient should be treated according to guidelines, depending on the therapeutic situation.

5.4 Endovascular therapy (EVT)

Advances in endovascular therapy (EVT) have been remarkable, especially in devices. In the iliac artery region, bare metal stents are commonly implanted and have good long-term results. However, stent grafts are often used in highly calcified lesions or in areas with long occluded lesions, and results are comparable to bare metal stents [16]. Surgery is the first choice for occlusion or stenosis of the common femoral artery. EVT is recommended only for patients at high surgical risk, but the results are inferior to surgery [1, 17]. In the femoral below-the-knee artery region, the primary patency of drug-eluting stent was significantly higher (88%) than that of drug-coated stent in a non-inferiority trial between drug-eluting stent and drug-coated stent due to the advancement of drug technology [18]. The JET STREAM® and Diamond back 360® are debulking devices that directly ablate plaques and calcified lesions, while the intra- and intra-eluting stents are debulking devices that directly ablate plaques and calcified lesions. Shock wave intravascular lithotripsy is a device that directly ablates plaques and calcified lesions, while Shock wave intravascular lithotripsy can fully dilate calcified vessels with balloon dilation by applying shock waves from the balloon. In addition, covered stents and interwoven stents are also available, and both work well in TASC C and D lesions, with good patency rates.

In the area below the knee and ankle joint, balloon angioplasty is the main treatment, but coronary stent implantation may be effective in some cases. In CLTI, the amputation avoidance rate was 93% at 3 years with stenting of 10 cm [19]. In addition, below the knee artery and ankle joint, asymptomatic patients should not be treated, and intervention should be considered if the patient has a Rutherford classification (**Table 1**) of 4 or higher.

The use of imaging modalities such as intravascular ultrasound (IVUS) and optical frequency domain imaging (OFDI) for peripheral vascular therapy is effective (**Figures 6** and 7). Although angiography is generally used for treatment, IVUS allows detailed measurement of the reference vessel diameter and selection of a suitable stent or balloon size [20]. Compared to IVUS, OFDI provides more detailed observation of vessel characteristics, but unlike IVUS, it requires the removal of blood cells during imaging, making it difficult to position the device implantation site in real time.

Fontaine classification		Rutherford's classification		
Grade	**Symptoms**	**Grade**	**Category**	**Symptoms**
I	Asymptomatic	0	0	Asymptomatic
IIa	Mild claudication	I	1	Mild claudication
IIb	Moderate or severe claudication	I	2	Moderate claudication
		I	3	Severe claudication
III	Ischemic rest pain	II	4	Ischemic rest pain
IV	Ulceration or gangrene	III	5	Minor tissue loss
		III	6	Major tissue loss

Table 1.
Lower extremity artery diseases—Fontaine and Rutherford's classifications.

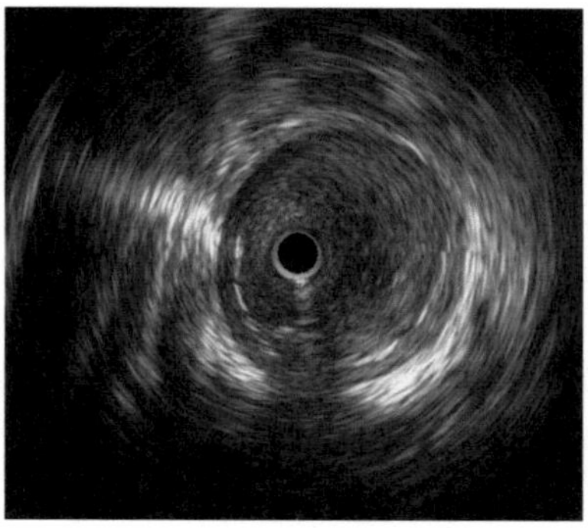

Figure 6.
Image of IVUS-courtesy of the author.

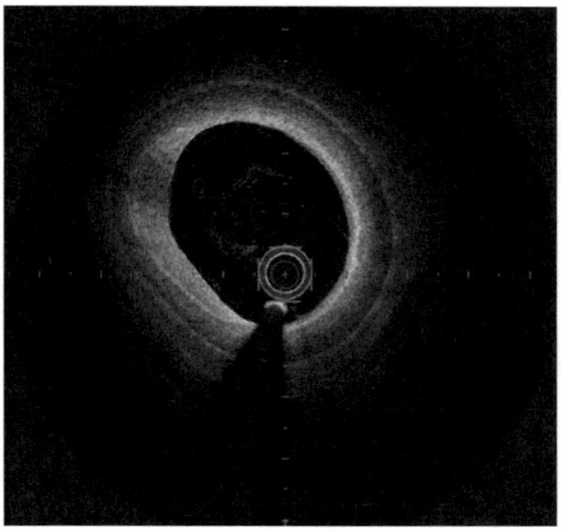

Figure 7.
Image of OFDI-courtesy of author.

The primary goal of endovascular or surgical intervention for LEAD is symptomatic improvement or wound healing.

6. Pathogenesis of LEAD

A primary factor in initiating the development of atherosclerotic lesions in the arterial system is vascular endothelial dysfunction. It is a complex and multifactorial process that results from natural changes in the function of the endothelium due to aging, as well as from adverse environmental factors and factors related to the metabolic status of the system. While there is no escaping the fact that the development of atherosclerosis is inevitable, it can be slowed down by appropriate lifestyle, diet, and medication choices. Increasing the antioxidant potential of endothelial cells is believed to delay aging, although the results of clinical studies are often contradictory. Elevated arginase I activity in endothelial cells accelerates their senescence by uncoupling endothelial nitric oxide (NO) synthase subunits with concomitant oxidative stress and inflammatory response. These processes were inhibited in the presence of N-acetylcysteine (NAC), a substrate of glutathione synthesis. Maintaining adequate endothelial NO production and its availability within the vascular bed is an important approach to slowing the development of atherosclerotic lesions. Another approach is to supplement with L-arginine, but also with L-citrulline, to slow down the aging of the endothelium, to inhibit oxidative stress, and to increase the production of NO. Administration of endothelial cell growth factors to stimulate new vessel formation or gene transfer to stimulate systemic production of these substances is another approach to treating or correcting atherosclerotic lesions. The applicability of autologous stem cells or endothelial progenitor cells for in vivo vascular reconstruction is also being investigated. Preliminary clinical trial results suggest the efficacy of these novel therapies. Preliminary clinical trial results indicate the effectiveness of these novel therapies in reducing pain and accelerating ulcer healing. In addition to these studies, research is underway to create biocompatible substrates that will facilitate the implantation and growth of stem cells to more rapidly initiate the formation of new blood vessels.

7. Microcirculation

The microcirculation constitutes the terminal network of the systemic circulation and encompasses micro vessels with diameters less than 20 μm, including arterioles, post-capillary venules, capillaries, and their associated cellular components. It represents the ultimate destination of the cardiovascular system, where the transfer of oxygen from the red blood cells in the capillaries to the parenchymal cells occurs, providing the necessary energy to sustain their functional activity. The microcirculation is also involved in the regulation of solute exchange between the intravascular and interstitial spaces, and facilitates the transport of hormones, nutrients, and immune cells to the tissue cells. As it is in direct contact with the parenchymal cells, the proper function of the microcirculation is vital for maintaining the viability of the organs and supporting their functions [21].

Endovascular therapy procedures aimed at reestablishing arterial perfusion to the targeted limb, such as percutaneous transluminal angioplasty, can significantly enhance transcutaneous oxygen tension in the revascularized distal limb. Patients who underwent amputation post-angioplasty exhibited no changes in transcutaneous oxygen tension, while those who displayed a progressive increase in the following weeks exhibited 100% limb salvage, implying that successful preservation of limb function necessitates the delivery of blood flow to the at-risk distal tissue. Microvascular dysfunction is characterized by a disturbance in the regulation of blood flow and vascular tone, leading to a decline in the delivery of oxygen to tissues, heightened oxidative stress, and a reduction in capillary density. Decreased microvascular density in calf muscle more accurately predicts lower limb functionality than the extent of atherosclerotic disease. The decline in microvascular density may serve as a common mechanism for limb dysfunction in various disease conditions that restrict leg function. The reduction of microvascular density in the skeletal muscle of the lower extremity also significantly contributes to the decline in exercise tolerance observed in patients with congestive heart failure [22].

In the coronary artery, coronary blood flow reserve (CFR) is an indication of the ability of the coronary circulation to deliver a maximal hyperemic blood flow and is indicative of impaired coronary microvascular function. Some patients with acute myocardial infarction (AMI) fail to achieve adequate tissue perfusion despite adequate epicardial blood flow after coronary intervention. Impaired CFR has been associated with microvascular injury resulting in increased morbidity and mortality [23].

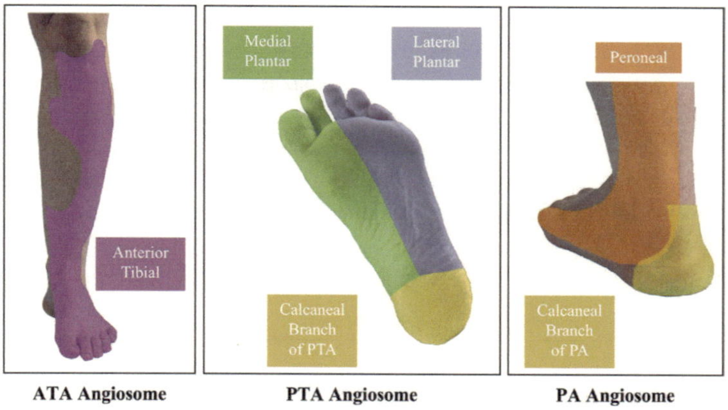

Figure 8.
Image of angiosome.

A similar attempt was made in LEAD, where Fukunaga et al. performed a physiological assessment in all the patients before and after EVT. The subjects were Rutherford classification 5 patients with ulceration or gangrene in the territory of the infrapopliteal artery. Physiological measurements were performed in the infrapopliteal artery region to measure vascular flow reverse (VFR). The wound-healed group had significantly higher VFR after EVT, while the wound-healed group had significantly higher VFR after EVT. Skin perfusion pressure (SPP) is currently considered useful in the evaluation of microcirculation [24], and SPP > 40 mmHg is associated with wound healing. However, there was no difference in mean SPP values between the two groups, and patients in the nonhealing group could not dilate their microvascular resistant arteries during hyperemia [25].

Studies have posited the value of the angiosome concept in the context of bypass surgery and endovascular procedures aimed at limb salvage (**Figure 8**) [26, 27]. The angiosomes of the foot and ankle are demarcated by the three main supplying arteries, namely the anterior tibial artery, posterior tibial artery, and peroneal artery. The hypothesis of direct blood flow to the ulcer or gangrene being influenced by the angiosome concept was analyzed.

The microcirculatory status was assessed by angiography, utilizing the wound blush method, following endovascular therapy. A positive result was manifested by either the blushing of the tissue surrounding the wound or the opacification of vessels around the wound by contrast. The presence of a wound blush following endovascular procedures is indicative of an increase in skin perfusion pressure, which is in turn associated with a higher likelihood of limb salvage. The utilization of wound blush as an angiographic endpoint during endovascular procedures may be a novel predictor of limb preservation outcomes in patients with critical limb ischemia [28].

8. Conclusions

The primary goals of LEAD treatment are wound healing and amputation prevention. Vascular surgeons and endovascular interventionalists have focused on improving blood flow to the ischemic limb and have developed and refined a variety of treatment modalities. However, in rare cases, improving blood flow does not improve tissue perfusion, which is exacerbated by the intervention. In these cases, microcirculation, which is not allowed in DSA, must be considered. As mentioned above, blood vessels have endothelial function and are involved in a variety of chemical mediators and vasa vasorum, and in LEAD patients, inadequate angiogenesis and collateralization may be a mechanism that increases limb ischemia and leads to dysfunction. This may be a mechanism for causing functional impairment [29].

Microcirculation evaluation is not stated in the guidelines, but it clearly contributes to cutaneous wound healing and plays an important role. There is a need for further research.

Conflict of interest

The authors declare no conflict of interest.

Author details

Yuki Matsumoto
Iwate Prefectural Ofunato Hospital, Ofunato, Japan

*Address all correspondence to: ma2pac625@gmail.com

References

[1] Aboyans V, Ricco JB, et al. 2017 ESC guidelines on the diagnosis and treatment of peripheral arterial diseases, in collaboration with the European Society for Vascular Surgery (ESVS). European Heart Journal. 2018;**39**(9): 763-816. DOI: 10.1093/eurheartj/ehx095

[2] Diehm C, Schuster A, et al. High prevalence of peripheral arterial disease and co-morbidity in 6880 primary care patients: Cross-sectional study. Atherosclerosis. 2004;**172**(1):95-105. DOI: 10.1016/S0021-9150(03)00204-1

[3] Fowkes FGR, Aboyans V, et al. Peripheral artery disease: Epidemiology and global perspectives. Nature Reviews Cardiology. 2017;**14**(3):156-170. DOI: 10.1038/nrcardio.2016.179

[4] Jawień A, Filipiak KJ, et al. Polskie wytyczne postępowania w chorobie tętnic kończyn dolnych (LEAD) na podstawie wytycznych ESVS/ESC 2017 - stanowisko ekspertów polskiego towarzystwa chirurgii naczyniowej, polskiego towarzystwa nadciśnienia tętniczego oraz sekcji farmakoterapii sercowo-naczyniowej polskiego towarzystwa kardiologicznego. Acta Angiologica. 2019;**25**(4):1-53. DOI: 10.5603/AA.2019.0015

[5] Norgren L, Hiatt WR, Dormandy JA, et al. TASC II Working Group. Inter-Society Consensus for the Management of Peripheral Arterial Disease (TASC II). European Journal of Vascular Endovascular Surgery. 2007;**35**(1):S1-S75. DOI: 10.1016/j.ejvs.2006.09.024

[6] Newman AB, Siscovick DS, Manolio TA, et al. Ankle-arm index as a marker of atherosclerosis in the Cardiovascular Health Study. Cardiovascular Heart Study (CHS) Collaborative Research Group. Circulation. 1993;**88**(3):837-845. DOI: 10.1161/01.cir.88.3.837

[7] Meijer WT, Grobbee DE, Hunink MG, et al. Determinants of peripheral arterial disease in the elderly: The Rotterdam study. Archieves in Internal Medicine. 2000;**160**(19):2934-2938. DOI: 10.1001/archinte.160.19.2934

[8] Allison MA, Criqui MH, McClelland RL, et al. The effect of novel cardiovascular risk factors on the ethnic-specific odds for pe ripheral arterial disease in the Multi-Ethnic Study of Atherosclerosis (MESA). Journal of American College of Cardiology. 2006;**48**(6):1190-1197. DOI: 10.1016/j.jacc.2006.05.049

[9] Criqui MH, Aboyans V. Epidemiology of peripheral artery disease. Circulatory Research. 2015;**116**(9):1509-1526. DOI: 10.1161/CIRCRESAHA.116.303849

[10] Joosten MM, Pai JK, Bertoia ML, et al. Associations between conventional cardiovascular risk factors and risk of peripheral artery disease in men. JAMA. 2012;**308**(16):1660-1667. DOI: 10.1001/jama.2012.13415

[11] Jude EB, Oyibo SO, Chalmers N, et al. Peripheral arterial disease in diabetic and nondiabetic patients: A comparison of severity and outcome. Diabetes Care. 2001;**24**(8):1433-1437. DOI: 10.2337/diacare.24.8.1433

[12] Ridker PM, Stampfer MJ, Rifai N. Novel risk factors for systemic atherosclerosis: A comparison of C-reactive protein, fibrinogen, homocysteine, lipoprotein(a), and standard cholesterol screening as predictors of peripheral arterial disease.

JAMA. 2001;**285**(19):2481-2485. DOI: 10.1001/jama.285.19.2481

[13] Rose S. Noninvasive vascular laboratory for evaluation of peripheral arterial occlusive disease: Part I —Clinical applications: Chronic, usually atherosclerotic, lower extremity ischemia. Journal of Vascular and Interventional Radiology. 2000;**11**(10):1257-1275. DOI: 10.1016/s1051-0443(07)61300-1

[14] Ouwendijk R, Kock MC, van Dijk LC, et al. Vessel wall calcifications at multi-detector row CT angiography in patients with peripheral arterial disease: Effect on clinical utility and clinical predictors. Radiology. 2006;**241**(2):603-608. DOI: 10.1148/radiol.2412050781

[15] Lim SS, Vos T, et al. A comparative risk assessment of burden of disease and injury attributable to 67 risk factors and risk factor clusters in 21 regions, 1990-2010: A systematic analysis for the Global Burden of Disease Study 2010. Lancet. 2012;**380**:2224-2260

[16] Piazza M, Squizzato F, et al. Editor's choice—Outcomes of self expanding PTFE covered stent versus bare metal stent for chronic iliac artery occlusion in matched cohorts using propensity score modelling. European Journal of Vascular and Endovascular Surgery. 2017;**54**(2):177-185. DOI: 10.1016/j.ejvs.2017.03.019

[17] Nakama T et al. 1-year outcomes of thromboendarterectomy vs endovascular therapy for common femoral artery lesions. JACC: Cardiovascular Interventions. 2022;**15**(14):1453-1463. DOI: 10.1016/j.jcin.2022.03.010

[18] Gray WA, Keirse K, et al. A polymer-coated, paclitaxel-eluting stent (Eluvia) versus a polymer-free, paclitaxel-coated stent (Zilver PTX) for endovascular femoropopliteal intervention (IMPERIAL): A randomised, non-inferiority trial. The Lancet. 2018;**392**(10157):1541-1551. DOI: 10.1016/S0140-6736(18)32262-1

[19] Giaquinta A, Vincenzo A, et al. Everolimus-eluting stent for patients with critical limb ischemia and Infrapopliteal arterial occlusive disease. Vascular and Endovascular Surgery. 2017;**51**(2):60-66. DOI: 10.1177/1538574416689429

[20] Iida O, Takahara M, et al. Vessel diameter evaluated by intravascular ultrasound versus angiography. Journal of Endovascular Therapy. 2022;**29**(3):343-349. DOI: 10.1177/15266028211047946

[21] Guven G, Hilty MP, Ince C. Microcirculation: Physiology, pathophysiology, and clinical application. Blood Purification. 2020;**49**(1-2):143-150. DOI: 10.1159/000503775

[22] Martens RJH et al. Microvascular endothelial dysfunction is associated with albuminuria: The Maastricht study. Journal of Hypertension. 2018;**36**(5):1178-1187. DOI: 10.1097/HJH.0000000000001674

[23] Taqueti VR, di Carli MF. Coronary microvascular disease pathogenic mechanisms and therapeutic options: JACC state-of-the-art review. Journal of the American College of Cardiology. 2018;**72**(21):2625-2641. DOI: 10.1016/j.jacc.2018.09.042

[24] Castronuovo JJ, Adera HM, et al. Skin perfusion pressure measurement is valuable in the diagnosis of critical limb ischemia. Journal of Vascular Surgery. 1997;**26**(4):629-637. DOI: 10.1016/S0741-5214(97)70062-4

[25] Fukunaga M, Fujii K, et al. Vascular flow reserve immediately after

infrapopliteal intervention as a predictor of wound healing in patients with foot tissue loss. Circulation: Cardiovascular Interventions. 2015;**8**(6). DOI: 10.1161/CIRCINTERVENTIONS.115.002412

[26] Attinger CE, Evans KK, Bulan E, et al. Angiosomes of the foot and ankle and clinical implications for limb salvage: Reconstruction, incisions, and revascularization. Plastic Reconstructive Surgery. 2006;**117**:261S-293S

[27] Iida O, Soga Y, et al. Long-term results of direct and indirect endovascular revascularization based on the angiosome concept in patients with critical limb ischemia presenting with isolated below-the-knee lesions. Journal of Vascular Surgery. 2012;**55**(2):363-370. DOI: 10.1016/j.jvs.2011.08.014

[28] Utsunomiya M, Nakamura M, et al. Impact of wound blush as an angiographic end point of endovascular therapy for patients with critical limb ischemia. Journal of Vascular Surgery. 2012;**55**(1):113-121. DOI: 10.1016/j.jvs.2011.08.001

[29] Hamburg NM, Creager MA. Pathophysiology of intermittent claudication in peripheral artery disease. Circulation Journal. 2017;**81**(3):281-289. DOI: 10.1253/circj.CJ-16-1286

Chapter 4

Evolving Paradigms in Laboratory Biomarkers of Fibrinolysis Phenotypes and Association with Post-Traumatic Vascular Thrombosis

Gordon Ogweno and Edwin Kimathi Murungi

Abstract

Traumatic tissue injury triggers blood coagulation to stanch bleeding and concomitant blood clot lysis to restore vascular patency. Approximately, 40% of trauma cases potentially present with trauma-induced coagulopathy that may coexist with clot dissolution or fibrinolysis. Laboratory test results of fibrinolysis biomarkers stratify fibrinolytic phenotypes into hyperfibrinolysis, physiological, hypofibrinolysis, and fibrinolytic shutdown. However, often, there is incongruence between laboratory findings and clinical presentation of bleeding or vascular thrombosis. Increasingly, it is becoming clear that laboratory findings transiently depend on the timing of blood sampling. The spectrum of evolving fibrinolysis phenotypes, a component of nature's adaptation to wound healing that ranges from initial promotion of blood fluidity to subsequent thrombosis, presents a clinical diagnostic dilemma with regard to the timing of antifibrinolytics or anticoagulants intervention. This chapter will review the available literature on post-traumatic fibrinolytic phenotypes, diagnostic challenges, evolution over time, clinical outcomes following therapeutic interventions, and association with vascular thrombosis.

Keywords: trauma, fibrinolysis phenotypes, therapeutic interventions, vascular thrombosis, laboratory biomarkers

1. Introduction

1.1 Clinical presentation of fibrinolysis in trauma

Post-trauma fibrinolysis is one of the causes of uncontrolled diffuse bleeding from injured areas despite surgical repair. The tissues rich in fibrinolytic activity are commonly urothelium such as the prostate, tonsils, uterus and placenta, kidneys, and urogenital system [1]. There is failure to stabilize forming blood clots, or alternatively, clot breakdown at level of the capillary bed resulting in diffuse bleeding involving

IntechOpen

uninjured sites and is extremely difficult to stop with mechanical interventions. Also, oozing from puncture sites and operative sites, excessive bruising, petechial hemorrhages, and expanding hematomas are not attributable to major vascular injury. The defining characteristics are generalized diffuse oozing from wounds, clot formation followed by dissolution, delayed bleeding post trauma, surgery, or procedures such as dental extractions.

Early observations on the fibrinolytic activity were in cadaveric blood obtained immediately from dead animals or victims of sudden violent death where blood remained liquid without clotting, inhibited clotting of plasma from normal living donors when mixed, or caused dissolution of clots formed from pure thrombin and fibrinogen. Others observed that plasma during surgery of certain organs formed clots in normal way but soon dissolved, unlike samples collected before operation that formed clots that remained solid for indefinitely for weeks.

1.2 Physiology of fibrinolysis

Fibrinolysis occurs when circulating inactive plasminogen is converted to active plasmin that breaks down fibrin to fibrin soluble products. The activator for plasminogen is tissue plasminogen activator (t-PA) released from injured endothelium or urokinase plasminogen activator (u-PA) from urothelium. Other activators of the fibrinolytic system, though playing minor roles under physiological conditions, are lipoproteins, Kallikrein-kinin system of the contact pathway of intrinsic coagulation [2].

The system is regulated by limiting the reaction complexes on fibrin or cellular surfaces [3, 4]. When bound to cross-linked fibrin, t-PA, plasminogen, and fibrin, they form a trimolecular complex, but while on endothelial surface, they form a tetramer made of annexin-2 as the linker to cellular surface and S100A10 as the receptor for plasminogen in addition to t-PA and plasminogen. On urothelium, urokinase substitutes for t-PA and is anchored to the surface by urokinase plasminogen activator receptor (u-PAR) [2]. In this arrangement, Annexin-2 and u-PAR act as cofactors to impede inactivation of surface-bound plasmin from circulating inhibitors. In solution or liquid phase plasminogen activator inhibitor- 1&- 2 (PAI-1 &2) complexes with t-PA to limit its interaction with fibrin, α2-antiplasmin (α2-AP) to inhibit active free plasmin to prevent indiscriminate action in circulation [2].

1.3 Activation of fibrinolysis In trauma

Fibrinolysis is activated in trauma by-products of tissue injury, development of hypo perfusion or shock state, and occlusive thrombi [5].

1.3.1 Relation to markers of tissue injury

Increased fibrinolysis is positively correlated with injury severity and endothelial injury markers such as syndecan-1 [6, 7].

1.3.2 Relation to shock state

Fibrinolysis is associated with and correlates with severity of shock states [8, 9]. Due to the induction of metabolic acidosis in shock and the fact that *in vitro* acidification precipitates euglobulin fractions [10, 11] and that acidification inactivates PAI-1 in solution [12], it is presumed that low pH causes fibrinolysis. However, this position is not in tandem with other empirical observations since such as (a) plasmin fibrinolytic activity is optimal at neutral pH and activity decreases at low pH [10, 13–17], (b) paradoxically, *in vivo* experimental infusion of lactic acid in laboratory animals was associated with microvascular thrombosis [18]. The discrepancy between the two positions can be accounted for by the fact that although acidosis diminishes plasminogen activator inhibitor -1 (PAI-1) activity, while enhancing the activators *in vitro*, this effect is minimal *in vivo* t since it is counterbalanced by metabolomics produced during anaerobic metabolism in shock that promote broad fibrinolysis in overall [19–21].

2. Laboratory measurement of fibrinolysis activity

2.1 Biomarkers of fibrin degradation

Plasmin sequentially degrades fibrin into several degradation products varying in molecular weights that have been quantified as evidence of fibrinolysis [22, 23].

D-dimers (DD), the terminal degradation products of ligated fibrin, are considered an indirect evidence of fibrin formation and its subsequent lysis [24]. The commonest method of measurement is ELISA-based, and several assays are available giving results in D-dimer units (DDU) or fibrinogen equivalent units (FEU). It is commonly used for exclusion of DVT with sensitivity of 96–100% and specificity of 40–70% [24]. However, its utility as a measure of plasmin fibrinolysis activity post trauma has come into question due to (a) non-plasmin sources such as from neutrophil elastase (NE) [25, 26] and (b) high variability owing to altered hemodynamic effects on hepatic clearance [27].

2.2 Assay of individual protein components of activators and inhibitors

Levels of proteins acting as activators and inhibitors of the fibrinolytic system are usually measured using immunological techniques, using free and complexed antigen, and enzyme chromogenic assays for activity levels in plasma. The quantity of protein levels varies with sex, age and diurnal variation, and status of catecholamine activation. Trauma increases the concentration and half-lives of these proteins [28]. Amongst the clinically available tests are for t-PA, PAI-1, PAI-2, alpha-2 antiplasmin (α2-AP), and the plasmin-antiplasmin (PAP) complex. Notably, plasmin levels are difficult to measure due to rapid inactivation in plasma [28].

2.3 Plasma global functional assays

Euglobulin clot lysis test (ECLT): This involves separation of platelet-free plasma (PFP), dilution with distilled water, and precipitation of globulin fraction to obtain euglobulin that is then acidified using acetic acid, centrifuged, and precipitate obtained redissolved in buffer solution. The acidified precipitate is rich in t-PA but poor in inhibitors such as PAI1, α2-AP, and TAFI [29]. Clot formation is initiated by recalcification and addition of thrombin in a test tube. Fibrinolytic activity is determined by observing the time taken to clot dissolution. Because of its dependence on fibrinogen levels, this test is insensitive in low fibrinogen states [30].

Fibrin plate lysis test: In this method, developed by Astrup and Mullertz [31], thrombin is added to fibrinogen solution following which euglobulin fraction, containing residual fibrin FXII and residual plasminogen, is added to start the reaction. The fibrinolysis is determined by measuring the diameter of lytic area or photometrically in a microplate reader [30]. This method is of historical importance and is currently rarely used currently due to its lengthy nature.

2.4 Whole blood viscoelastic hemostatic assays (VHA)

The viscoelastic platforms detect changes in viscosity of blood against the wall in a rotating pin and cup after coagulation has been triggered. While thromboelastolastography (TEG) or thrombelastometry (ROTEM) of the commercially available instruments are commonly deployed in trauma, Sonoclot has not found wide application. Commercial instruments for these assays differ in mechanics, triggering agents, and nomenclature of machine readouts [29]. These methods are hugely popular in the point-of-care diagnosis of active and ongoing fibrinolysis, unlike the molecular assays that are time-consuming and are dependent on plasma clearance and give indication of dynamic changes in blood. TEG and ROTEM display graphically the evolution against time (onset of clot formation, rate of fibrin build-up, development of maximum clot strength, and thereafter its reduction indicating clot dissolution) [32]. In either of the systems, fibrinolysis is indicated by the percentage reduction in area under the curve as lysis % (Ly %) assuming MA remains constant) on TEG, or the percentage of clot remaining after MCF as Lysis index % (LI %), or as maximum lysis % (ML%) on ROTEM at 30 minutes and 60 minutes [33]. For fibrinolytic phenotypes stratification, while the cutoff for TEG parameters has been classified as hyperfibrinolysis (Ly 30 > 3%), physiologic (Ly 30 0.81–2.9%), and shutdown (Ly 30 < 0.8%) [34]; the cutoff for ROTEM parameters are physiologic (ML 3–15%), shutdown (ML <3%), and hyperfibrinolysis (ML > 15%) [35]. When a particular pattern of severe fibrinolysis termed 'death diamond' is present, the likelihood of hemorrhage and mortality is high [36, 37].

ROTEM has been utilized to demonstrate high rates of hyperfibrinolysis in trauma patients [38]. Due to the increased incidences of hypofibrinolysis or fibrinolytic shutdown in trauma and surgery, the method has been modified to incorporate plasminogen activators to increase sensitivity [39]. Validation and standardization of ROTEM [40, 41] have shown good congruence with alternative methods for fibrinolysis measurements. However, great discrepancy has been reported between the VHA and molecular biomarkers of fibrinolysis in trauma [42].

3. Concepts in fibrinolysis

3.1 Local fibrinolysis

Fibrinolysis is initiated by localized tissue injury that leads to release of plasminogen activators in general circulation. Some tissues, such as endothelium, endometrium, tonsils, and urothelium, express high amounts of plasminogen activators [43], and therefore would be expected to exhibit localized clot lysis without systemic involvement. Indeed, surgical incisions or injury of these anatomical sites have been observed to evince characteristic fibrinolytic hemorrhage without systemic changes in biomarkers [44]. The process is aided by localized plasminogen receptor S100A10 and their endothelial receptor of Annexin-2 on vascular linings *in vivo* [45, 46]. Local fibrinolysis ensures vascular patency of vascular territories without affecting clots formed elsewhere, therefore, ensuring cardiovascular integrity [47].

3.2 Systemic fibrinolysis

In systemic fibrinolysis, there is evidence of fibrinolysis activity in the general circulation. Characteristic features include increased t-PA, PAP, and fibrin degradation products such as FDPs or D-dimers. Additionally, bleeding is generalized and not restricted to particular sites. It is often seen in DIC, severe trauma, circulatory hypoperfusion/ shock, or administration of thrombolytics [42, 48].

3.3 Primary fibrinolysis

This is the normal process that occurs during physiological conditions when there is optimal balance between the activators and the inhibitors that keep the vascular lumen patent [49]. Primary fibrinolysis implies a response arising in the absence of hypercoagulable or thrombotic state [50]. Usually, bleeding or thrombosis risks are only increased in case of congenital deficiency in one or more of the factors. It is rarely encountered in trauma.

3.4 Secondary fibrinolysis

This process is secondary to a pathological or disease process causing imbalance in the regulation of fibrinolysis process. Secondary fibrinolysis develops in response to intravascular thrombin generation and fibrin deposition [50]. The evident dysregulation could involve one or a combination of factors including alteration of rate of endothelial secretion, release of t-PA or alternative activators, or alteration of concentration of inhibitors such as PAI-1 and hepatic clearance of t-PA or other activators. Although many disease processes cause secondary fibrinolysis, in trauma tissue damage that is associated with shock and acidosis leads to t-PA secretion and delayed clearance that overwhelm the natural inhibitors leading to consequent hyperfibrinolysis. Furthermore, the procoagulant tissue factor that is released during tissue injury participates in initiation of coagulation, microthrombi formation initialization of cascade of events that lead to plasminogen activation.

3.5 Physiological fibrinolysis

The concept of physiological fibrinolysis is based on a balance between activators and inhibitors of fibrinolytic system leading to an equilibrium between activated and nonactivated states [51]. This is the ideal and naturally intended state during trauma response where vascular patency is maintained but without bleeding.

3.6 Pathological fibrinolysis

Although fibrinolysis is a protective mechanism geared toward to restoration of vascular patency, it can become excessive leading to hemorrhage or insufficient leading to thrombosis [50]. Majority fibrinolysis phenotypes in trauma belong to this group presenting as either hyperfibrinolysis or shutdown.

3.7 Fibrinolysis phenotypes in trauma

Using a combination of plasma protein assays and VHA, the fibrinolysis in trauma has been stratified into hyperfibrinolysis, physiological, and shutdown, reflecting outcomes related to organ failure or hemorrhage [34].

3.7.1 Hyperfibrinolysis

This is when fibrinolysis activation causes or sustains a bleeding tendency [50]. It is associated with severe injury, prolongation of routine coagulation tests, and greater mortality [38]. Moreover, it is the most lethal phenotype of lysis pattern in trauma resulting in high risks of hemorrhage and high mortality. The laboratory criteria characteristically show markedly elevated PAP and DD, as well as increased lysis on vHA [52]. The cause of in trauma is the markedly elevated t-PA that rapidly and massively activates plasminogen to plasmin overwhelming the inhibitory factors such as PAI-1 and alpha-2 antiplasmin [53, 54].

3.7.2 Physiological fibrinolysis

The laboratory evidence for physiological fibrinolysis includes a combination of normal or slightly elevated plasma PAP and DD together with normal or low lysis on VHA [52]. Although this phenotype has low mortality rates, patients usually require blood transfusion, suffer organ failure, or hemorrhage [34].

3.7.3 Occult fibrinolysis

In this, patients who have concealed evidence of fibrinolysis [55]. Characteristically, despite low lytic activity on VHA (low ROTEM %ML) patients exhibit paradoxically, discordant high plasma levels of fibrinolysis biomarkers such as (DD) and S100A10. However, other biomarkers such as t-PA, plasmin-anti plasmin (PAP), PAI-1, and PF 1+2 remain normal or not elevated. Furthermore, there is evidence of thrombin generation given the high F1+2 fibrinopeptide [46]. Globally, there is a balance between fibrinolysis, thrombin activation (PF 1&2), fibrin formation (FP A & FPB), and degradation (DD) such that coagulation tests (fibrinogen, PT, and aPTT) are normal despite elevated DD and low ML on ROTEM [42]. The discrepancy in test results is plausibly due to the high level of S100A10, which is active on cellular membrane surfaces *in vivo*, but without activity *ex vivo* under testing conditions. The membrane-bound S100A10, plasminogen, t-PA, and Annexin-2 bind onto cell surfaces in close proximity form a tetramer. It is upregulation under hypoxic conditions in trauma leads to localized fibrin degradation despite low systemic t-PA-plasmin levels.

3.7.4 Lowfibrinolysis

3.7.4.1 Hypofibrinolysis

This condition occurs due to lowfibrinolytic state without evidence of prior activation due to impaired t-PA release in the presence of elevated baseline fibrinolytic inhibitors [56]. The terminology deployed in this case is contentious given that it is also considered to correspond to a failure of fibrinolysis activation following clotting, especially when ELT is the test [29]. Traditionally, these hypofibrinolysis patients were taken to be those lacking fibrinolytic activity on venous occlusion test. However, this test is now largely obsolete and impractical in trauma victims, though the test has fallen out of favor and may not be practical in trauma victims. The laboratory features include normal routine coagulation tests, lowfibrinolytic activity on VHA, but DD and PAP levels are proportional to injury burden. A unique feature of this phenotype is that patients are responsive to recombinant t-PA challenge unlike in fibrinolysis [56].

3.7.4.2 Fibrinolysis resistance

In fibrinolysis resistance, which potentially may also include hypofibrinolysis and fibrinolysis shutdown on ECLT test [29], formed blood clots or thrombi are resistant to lysis by plasminogen activators [57, 58], and thus persists in vascular lumen [59]. In plasma, resistance is quantified by VHA (TEG/ROTEM) modified assays spiked with t-PA [60] in which resistant samples are unresponsive. Fibrinolysis resistance, putatively due to an imbalance between endogenous t-PA and PAI-1, shares some similarities with fibrinolysis shutdown, although other factors such as fibrin architecture, vWF, NETs, and microparticles are thought to offer *in vivo* contribution [58].

3.7.4.3 Fibrinolysis shutdown

In this condition, it is postulated that fibrinolysis, activated at some time point, is inactivated by the time of blood sampling. This phenotype was decoded over 60 years ago [61, 62] after observing that dilute blood lysis time and ELT were shorter 45 minutes into surgical operations but became prolonged during the postoperative period consistent with decreased fibrinolysis. The long lysis time on ELT correlated with reduced t-PA activity with concurrent elevated peak t-PAI-1 inhibitor on the first postoperative day patterns, which were similar to myocardial infarction and trauma at diagnosis and during recovery phases [63, 64].

Prior to adoption of VHA in surgery and trauma, ELT showed lack of response on subsequent testing of the same patient after some time interval. The adoption of VHA helped to explain the discrepancy in the test findings due to rebound PAI-1 inhibitors after triggering of coagulation and release of t-PA [29]. Currently, fibrinolysis shutdown is demonstrated by elevated DD and PAP but decreased functional lysis on VHA (either TEG or ROTEM) [52]. Two tests, t-PA hypersensitivity rapid TEG (r-TEG) or r-ROTEM, are used to delineate lowfibrinolysis due to hypofibrilysis from shutdown [65]. Fibrinolysis shutdown is associated with high rates of organ failure and death [34].

The postulated mechanisms of fibrinolysis shutdown include (a) PAI-1 inhibition, (b) neutrophil elastase impairment of fibrinolysis [66] through degradation of plasminogen [67] and plasmin [68], (c) and degradation of alpha-chain of fibrin that blocks activation of plasminogen [69].

There are variants of fibrinolysis shutdown that include:

a. Acute: There is evidence of fibrinolysis activation such as elevated DD and PAP, and low t-PA due to rapid hepatic clearance within 1 hour of trauma. In addition, VHA displays lowfibrinolytic activity unrelated to inhibitors since they have been depleted through consumption but retain increased sensitivity to t-PA challenge [56].

b. Acquired: Early descriptions acquired fibrinolysis was in association with the administration of exogenous thrombolytics that became resistant to subsequent exposure to the same agent due to the development of inhibitors. In trauma, the onset of acquired fibrinolysis commences after 2 hours but can last for up to 24 hours. It is characterized by low fibrinolytic activity on VHA, low t-PA activity with increased resistance to lysis upon t-PA challenge, and an upregulation of PAI-1 [56].

c. Persistent: This is exhibited by fibrinolysis activation followed by delayed upregulation of inhibitors whose onset is after 24 hours and may persist for 7 days.

It is associated with platelet degranulation with persistent release of PAI-1 into systemic circulation signaling failure to regain hemostatic homeostasis [56]. The mechanism for persistent fibrinolysis shutdown is still unclear, but a strong immunological implicated. An alternative mechanism is the presence of neutrophil elastase (NE) released from leukocytes that not only degrades fibrin to DD but also digests plasminogen eventually contributing to fibrinolytic shutdown [67].

d. Fibrin resistant: An emerging phenomenon in cases of absence of fibrinolytic activity in trauma is the development of fibrin architecture resistant to plasmin lysis [70]. In this phenotype, viscoelastic hemostatic assays (TEG or ROTEM) demonstrate no lysis despite high circulating levels of t –PA released from injured sites and/or exogenous t-PA spiking during testing. The postulated mechanism is rapid and massive thrombin generation that leads to the development of denser and coarse fibrin networks impermeable by plasmin lytic head [71, 72]. In addition, it has been demonstrated that TAFI activation by thrombin inhibits plasminogen interaction with fibrin [73, 74], as well as FXIII cross-linking of fibrin confer lytic resistance [57].

3.8 Evolution of fibrinolysis phenotypes over time period

The incidence of the different fibrinolytic phenotypes varies and depends on the severity of injury, assays type employed and time period post-injury when blood was drawn for analysis. Notably, plasma fibrinolytic activity is a dynamic process in which transition from one phenotype to another frequently occurs.

In trauma patients, levels of molecular protein biomarkers levels change with time. It has been reported that while t-PA levels drastically decrease in the first hour and continue to decrease over time, PAI-1 levels incrementally increase for five hours post trauma in an inverse manner to t-PA [75]. However, a DD pattern is unrelated to either, since there was abrupt increase over first hour and then rapid drop in levels. The events after one hour probably reflect the fibrinolytic shutdown.

Serial determination of euglobulin lysis time (ELT) or fibrin platelet method in surgical patients has revealed that the period of peak enhanced fibrinolysis activity coincides with mid surgery, followed by progressive shutdown occurring over the next 24 hours post-operation, and thereafter leveling off by day third or fourth days [76]. This reflects the dynamic nature of fibrinolytic actors and measurements taken at single particular time points may be less useful as opposed to the trend in assay results.

By far VHA (TEG or ROTEM) has been used in trauma patients to stratify in fibrinolytic phenotypes in trauma patients. At admission, a majority (44–64%) of patients are in fibrinolytic shutdown, which persists for 120 hours [77]. The first 3 hours are critical since it is the time of transition from hyperfibrinolysis to physiologic state that remains stable for the next 5 days. The hyperfibrinolysis and physiological phenotypes may be described as transitory since they diminish within a few hours, while shutdown increases (from 27–60%) over three hours from the time of trauma [78]. As such, over time, shutdown is the predominant phenotype. Patients who are in initial hypofibrinolysis/shutdown and who remain in the same status or who transition from or to other phenotypes are prone to high mortality, especially due to multiple organ failure [79]. Transitions from one fibrinolysis phenotype to another may be due to resuscitation measures, rapid hepatic clearance of activators, and administration of antifibrinolytics [56]. In liver transplant recipients, initial hyperfibrinolysis eventually transition to either early or late shutdown over 120 minutes [80, 81].

4. Post-trauma fibrinolysis and development of vascular thrombosis

4.1 Epidemiology of vascular thrombosis post trauma

Cases of post-trauma vascular thrombosis are grossly under-reported owing to the difficulties in diagnosing the condition. Available evidence from clinical, radiological, histopathological, and autopsy studies that when present post-traumatic vascular thrombosis is preceded by fibrinolytic activity. Though varying depending on the vascular territory and patient population literature indicates that with surveillance, the rates of vascular could be up to 60% for DVT [82] and 25% for PTE [83], while microvascular thrombosis is variable [84, 85]. Although the macrovascular events are amenable to radiological visualization or imaging techniques, microvascular occlusions, especially at trauma sites are never included in the diagnostic reports. Furthermore, arterial thrombosis events including myocardial infarctions, ischemic stroke, or renal involvements are usually excluded from post-traumatic vascular thrombosis reporting, even though they occur frequently. Surprisingly, the incidence of post-trauma venous thrombosis is high despite pharmacological anticoagulant thromboprophylaxis [86]. The inadequate response to anticoagulants strongly suggests that fibrinolytic impairment superimposed on the coagulation cascade contributes to post-traumatic vascular.

4.2 Relation to fibrinolytic activity

Despite clinical and laboratory evidence of post-trauma fibrinolysis [42, 75], vascular thrombosis, in which impaired fibrinolysis plays a role, fibrinolysis [87, 88] remains a silent killer. There are three competing theories linking dysregulated fibrinolysis with thrombosis in trauma: (a) activation followed by suppression of lytic pathways leading to reduction of fibrin degradation products (FDPs) and D-dimers, (b) coexistent of lytic activity with overwhelming fibrin deposition associated with increased fibrin split products, and (c) formation of lysis resistant fibrin.

Fibrinolysis shutdown has largely been demonstrated in association with surgery [63] and has been shown to lead to high incidence of thrombosis complications despite heparin administration [89]. Many reports of thromboembolism in trauma have implicated fibrinolytic shutdown [80, 90].

Proponents of overwhelming fibrin deposition in the presence of active fibrinolysis have used D-dimers as a marker of thrombosis. In a cohort of trauma patients, serial D-dimer levels remained elevated above normal cutoff threshold levels for 14 days without evidence of bleeding. Instead, some patients were found to have PE. The elevated D-dimers though predictive of thrombosis outside trauma are difficult to interpret post-injury [91]. Similarly in surgery, postoperative fibrinolysis shutdown ascertained by ELT and fibrin plate method coincided with decreased FDP linked with the development of DVT or PE [76].

In non-trauma patients, vascular thrombosis has been linked to lowfibrinolysis states [88]. In such patients, hypofibrinolysis coincides with elevated inhibitor levels. This scenario is complicated by the concurrent increased t-PA in shutdown states. In effect, the elevated t-PA antigens are more predictive of thrombosis than fibrinolysis owing to their utility as a marker of endothelial damage [92]. This is because t-PA fibrinolytic activity is limited to plasminogen bound to cross-linked fibrin and surfaces, and therefore free-floating levels are insignificant to clot lysis.

Trauma patients are procoagulant [93] and show enhanced thrombin generation [94]. Thrombin contributes to fibrinolysis resistance by reducing profibrinolytic effects of thrombomodulin (TM) and activated protein C (APC) [73]. There is activation of TAFI, which blocks the binding of plasminogen to fibrin and activation of coagulation factor XIII, which covalently ligates fibrin, thus confers lysis resistance. Fibrin formed in the presence of high levels of thrombin is compact and impervious to permeation by plasmin lytic head [95, 96]. Furthermore, the formed thrombus in trauma is resistant to lysis due to incorporated RBC [97–99], FXIII [57], platelets [100, 101], and abnormal dense fibrin structure [59, 70–72, 102]. Other contributors to lysis resistance are neutrophil extracellular traps (NETs) [103–105] and circulating microparticles [106].

In trauma, the reactions of cellular elements are dynamic, with platelet reactions playing special role on lytic resistance. Specifically, in trauma, platelets contribute to lytic resistance by (a) producing over 90% of PAI-1, (b)release of FXIII that cross-links fibrin to diminish permeability by plasmin, (c) provision of a surface for catalytic coagulation reactions leading to thrombin burst and TAFI activation and release, (d) undergo clot retraction and release microparticles that promote coagulation cascade leading to overwhelming fibrin formation and deposition [107].

Tissue injuries associated with surgery, and trauma are associated with fibrinolysis in the early phases and vascular thrombosis in the later recovery phases. Although the fibrinolytic phenotypes evolve from one type to another, the shutdown phenotype increases with time [78] and is associated with the development of vascular thrombosis [80, 90, 108]. Although no single laboratory test can predict transition from fibrinolysis to thrombosis, some small studies indicate that resistance to t-PA on rapid thromboelastography/elastometry could be a useful test [109]. However, large clinical trials are needed to validate the claim.

5. Clinical experiences in surgery and trauma

5.1 Fibrinolysis and liver transplantation

Early in orthotopic liver transplantation hemorrhage due to fibrinolysis demonstrated by TEG was an issue and administration of tranexamic acid reduced requirements for blood transfusion [110]. With this evidence, empiric Epsilon amino caproic acid (EACA) antifibrinolytics was implemented. However, it was observed that over time that such practice increased vascular thrombosis [111]. Perhaps these early investigators did not appreciate transient time course of hyperfibrinolysis that occurs during anhepatic stage that soon resolves after re-establishment of circulation [112, 113]. Although all the fibrinolytic phenotypes have been observed during liver transplantation [81], but the shutdown phenotype is associated with the development of thrombohaemorrhagic complications that begins 30 minutes after re-establishment of reperfusion in allograft recipient [80] that is predicted by thromboelastography [114]. With this knowledge, it can be potentially concluded that indiscriminate administration of antifibrinolytics against hyperfibrinolysis if not followed by anticoagulants may increase risk of vascular thrombosis. In reviews of antifibrinolytics in liver transplantation over the last 4 decades, it is quite evident that the twin problems of hemorrhage secondary to hyperfibrinolysis and thrombosis still remain real challenges [115]. The clinical challenge is lack of a useful predictive test to guide timing of antifibrinolytics and when to start anticoagulants.

5.2 Fibrinolysis and cardiac surgery

Fibrinolysis is activated during cardiac surgery, especially so during cardiopulmonary bypass [116]. Antifibrinolytics are usually administered to decrease blood loss during cardiac surgery. However, the most potent agent aprotinin has been associated with significant thromboembolism and as a result, its indication has been discontinued in many countries [117]. Given that all fibrinolytic phenotypes have been described in cardiac surgery, and that the incidence of shutdown ranging from 30 to 45% [118], it is still controversial the contribution of aprotinin to fibrinolytic shutdown and vascular thrombosis. Furthermore, it is not yet known the proportion of patients who would transition from hyperfibrinolysis to shutdown and the relation of shutdown to thrombosis in such patients. What is clear is antifibrinolytics should be administered following a strict [119], even though simple laboratory guidelines are still lacking.

5.3 CRASH-2

The use of tranexamic acid (TXA) against fibrinolysis to forestall blood loss in surgery and trauma has been practiced since its introduction more than 60 years ago. Antifibrinolytic activity of TXA can only be monitored by VHA or DD changes but not by using molecular biomarkers such as t-PA, PAP, and PAI-1. In the largest randomized clinical trial of TXA in trauma (CRASH-2 trial), it was shown that, compared to placebo, TXA radically reduced hemorrhage and mortality. Furthermore, incidences of vascular thrombosis were not significant. In this study, TXA only had clear benefit if administered within 3 hours of trauma. However, other clinical trials post-CRASH-2 have shown that indiscriminate administration was associated with high proportions of fibrinolysis shutdown and increased risk of thrombosis by a factor almost twice than those that did not. This finding reinforced an earlier observation of increased thromboembolism in animals administered TXA to reverse hemorrhagic shock-induced hyperfibrinolysis. The current call for selective and rational usage in trauma is based on the recognition that it is associated with greater fibrinolytic shutdown but does not improve clot strength for those already in shutdown. A greater benefit is achieved if TXA is given within 3 hours post-injury (CRASH-2) and in those with proven hyperfibrinolysis phenotype who form less than 20% of cases. Furthermore, indiscriminate usage of tranexamic acid is associated with increased risk of development of post-traumatic venous thromboembolism. Although some studies have indicated that TEG Ly3% could be the critical cutoff to start TXA in trauma patients [120], systematic reviews of published results of TXA, and thrombotic events are still conflicting since some point to increased incidence [121], while others are unequivocal owing to heterogenous cohort of the patient population studied [122].

6. Conclusion

Traumatic injury is associated with concomitant activation of blood coagulation cascade and fibrinolysis. Pathological imbalance between the two could potentially lead to increased risk of hemorrhage or vascular thrombosis. No single laboratory test best describes the spectrum of fibrinolysis in trauma, and thus a combination is used to stratify fibrinolysis phenotypes. The phenotypes are increased of hyperfibrinolysis, normal or physiologic, and decreased fibrinolysis activity, which may be further categorized as hypofibrinolysis, fibrinolysis resistance, and fibrinolysis shutdown. The fibrinolytic phenotypes evolve over time and may transition from one type to another. Overall, the increased fibrinolysis activity is present in the early phases post trauma, and decreased activity or shutdown predominates in the later phases coinciding with recovery. Fibrinolysis shutdown has been associated with the development of occlusive vascular thrombosis. The evolution of fibrinolysis phenotypes is modified by resuscitation strategies in the management of hemodynamics associated with shocks such as intravenous fluids and blood product transfusions. Furthermore, therapeutic interventions such as antifibrinolytics may accelerate transition to fibrinolysis shutdown thereby increasing the risk of vascular thrombosis. The twin problems of post-traumatic fibrinolysis and thrombosis still remain a challenge, but unfortunately, laboratory guidelines in the management of therapeutic interventions to forestall hemorrhage from fibrinolysis and minimization of transition to thrombosis remain largely un described.

Author details

Gordon Ogweno[1*] and Edwin Kimathi Murungi[2]

1 Department of Medical Physiology, Kenyatta University, Kenya

2 Department of Medical Biochemistry, Kisii University, Kenya

*Address all correspondence to: ogweno.gordon@ku.ac.ke

References

[1] Tillett WS, Garner RL. The fibrinolytic activity of hemolytic streptococci. The Journal of Experimental Medicine. 1933;**58**(4): 485-502. DOI: 10.1084/jem.58.4.485

[2] Cesarman-Maus G, Hajjar KA. Molecular mechanisms of fibrinolysis. British Journal of Haematology. 2005;**129**(3):307-321. DOI: 10.1111/j.1365-2141.2005.05444.x

[3] Booth NA. Regulation of fibrinolytic activity by localization of inhibitors to fibrin(ogen). Fibrinolysis and Proteolysis. 2000;**14**(2-3, 213):206. DOI: 10.1054/fipr.2000.0071

[4] Urano T, Castellino FJ, Suzuki Y. Regulation of plasminogen activation on cell surfaces and fibrin. Journal of Thrombosis and Haemostasis. 2018;**16**(8):1487-1497. DOI: 10.1111/jth.14157

[5] Zipperle J et al. Conventional and pro-inflammatory pathways of fibrinolytic activation in non-traumatic Hyperfibrinolysis. Journal of Clinical Medicine. 2022;**11**(24):1-12. DOI: 10.3390/jcm11247305

[6] Johansson PI, Stensballe J, Rasmussen LS, Ostrowski SR. A high admission syndecan-1 level, a marker of endothelial glycocalyx degradation, is associated with inflammation, protein C depletion, fibrinolysis, and increased mortality in trauma patients. Annals of Surgery. 2011;**254**(2):194-200. DOI: 10.1097/SLA.0b013e318226113d

[7] Richter RP et al. Endotheliopathy is associated with a 24-hour fibrinolysis phenotype described by low TEG lysis and high d-dimer after trauma. Annals of Surgery Open. 2022;**3**(1):e116. DOI: 10.1097/as9.0000000000000116

[8] Cotton BA et al. Hyperfibrinolysis at admission is an uncommon but highly lethal event associated with shock and prehospital fluid administration. Journal of Trauma and Acute Care Surgery. 2012;**73**(2):365-370. DOI: 10.1097/TA.0b013e31825c1234

[9] Moore HB et al. Shock-induced systemic hyperfibrinolysis is attenuated by plasma-first resuscitation. Journal of Trauma and Acute Care Surgery. 2015;**79**(6):897-903. DOI: 10.1097/TA.0000000000000792

[10] Lauritsen OS. Inhibition of plasminogen activation and plasmin activity after euglobulin precipitation and acidification of plasma. Scandinavian Journal of Clinical and Laboratory Investigation. 1969;**23**(2):121-128. DOI: 10.3109/00365516909077014

[11] de Maat MPM, Kluft C, de Boer K, Knot EAR, Jie AFH. Acid treatment of plasma for the inactivation of plasminogen activator Inhiitor-1 *PAI-1. Thrombosis Research. 1988;**52**:425-430

[12] Kvassman JO, Lawrence DA, Shore JD. The acid stabilization of plasminogen activator inhibitor-1 depends on protonation of a single group that affects loop insertion into β-sheet A. The Journal of Biological Chemistry. 1995;**270**(46):27942-27947. DOI: 10.1074/jbc.270.46.27942

[13] Troll W, Sherry S, Wachman J. The action of plasmin on synthetic substrates. The Journal of Biological Chemistry. 1954;**208**(1):85-93. DOI: 10.2307/j.ctt7ztpn1.12

[14] Westlund LE, Andersson LO. Studies on the influence of reactants and buffer environment on clot lysis induced by human plasminogen activators. Thrombosis Research. 1985;**37**(1):213-223. DOI: 10.1016/0049-3848(85)90048-9

[15] Hedlin AM, Monkhouse FC. Fibrinolytic activities of Euglobulins precipitated at pH 6.4, 6.0, 5.3. Canadian Journal of Physiology and Pharmacology. 1969;**47**:935-940

[16] Gersel-Pedersen N. Fibrinolytic activity of salivary euglobulin fractions precipitated at pH 5.9 or 6.4 before and after mixing with blood. International Journal of Oral Surgery. 1980;**9**(3):190-197. DOI: 10.1016/S0300-9785(80)80018-4

[17] Hong SY, Shin HK, Yang DH. The effect of dilution, pH and ionic strength of plasma on t-PA precipitation in Euglobulin fraction. The Korean Journal of Internal Medicine. 1992;**7**(2):73-80

[18] Broersma RJ, Bullemer GD, Mammen EF. Acidosis induced disseminated intravascular microthrombosis and its dissolution by streptokinase. Thrombosis et Diathesis Haemorrhagica. 1970;**24**(1):55-67. DOI: 10.1055/s-0038-1654211

[19] Peltz ED et al. Pathologic metabolism: An exploratory study of the plasma metabolome of critical injury. Journal of Trauma and Acute Care Surgery. 2015;**78**(4):742-751. DOI: 10.1097/TA.0000000000000589

[20] Moore HB et al. Increase in post-reperfusion sensitivity to tissue plasminogen activator-mediated fibrinolysis during liver transplantation is associated with abnormal metabolic changes and increased blood product utilisation. Blood Transfusion. 2019;**17**(4):312-320. DOI: 10.2450/2019.0205-18

[21] Moore HB et al. Hemolysis exacerbates hyperfibrinolysis, whereas platelolysis shuts down fibrinolysis: Evolving concepts of the spectrum of fibrinolysis in response to severe injury. Shock. 2015;**43**(1):39-46. DOI: 10.1097/SHK.0000000000000245

[22] Connaghan DG, Francis CW, Labe DA, Marder VJ. Specific identification of fibrin polymers, fibrinogen degradation products, and crosslinked fibrin degradation products in plasma and serum with a new sensitive technique. Blood. 1985;**65**(3):589-597

[23] Eisenberg PR, Jaffe AS, Stump DC, Collen D, Bovill EG. Validity of enzyme-linked immunosorbent assays of cross-linked fibrin degradation products as a measure of clot lysis. Circulation. 1990;**82**(4):1159-1168. DOI: 10.1161/01.CIR.82.4.1159

[24] Johnson ED, Schell JC, Rodgers GM. The D-dimer assay. American Journal of Hematology. 2019;**94**(7):833-839. DOI: 10.1002/ajh.25482

[25] Seitz R, Lerch L, Immel A, Egbring R. D-dimer tests detect both plasmin and neutrophil elastase derived Split products. Annals of Clinical Biochemistry. 1995;**32**(2):193-195. DOI: 10.1177/000456329503200211

[26] Bach-Gansmo ET, Godal HC, Skjønsberg OH. Degradation of fibrinogen and cross-linked fibrin by human neutrophil elastase generates D-like fragments detected by ELISA but not latex D-dimer test. Thrombosis Research. 1998;**92**(3):125-134. DOI: 10.1016/S0049-3848(98)00121-2

[27] Chen JP, Rowe DW, Enderson BL. Contrasting post-traumatic serial

changes for D-dimer and PAI-1 in critically injured patients. Thrombosis Research. 1999;**94**(3):175-185. DOI: 10.1016/S0049-3848(98)00211-4

[28] Pepperell D, Morel-Kopp M-C, Ward C. Clinical application of fibrinolytic assays Dominic. In: Kolev K, editor. Fibrinolysis and Thrombolysis. Zagreb: IntechOpen; 2014. p. 13

[29] Longstaff C. Measuring fibrinolysis: From research to routine diagnostic assays. Journal of Thrombosis and Haemostasis. 2018;**16**(4):652-662. DOI: 10.1111/jth.13957

[30] Ilich A, Bokarev I, Key NS. Global assays of fibrinolysis. International Journal of Laboratory Hematology. 2017;**39**(5):441-447. DOI: 10.1111/ijlh.12688

[31] Astrup T, Mullertz S. The fibrin plate method for estimating activity. Archives of Biochemistry and Biophysics. 1952;**40**(2):346-351

[32] Hans GA, Besser MW. The place of viscoelastic testing in clinical practice. British Journal of Haematology. 2016;**173**(1):37-48. DOI: 10.1111/bjh.13930

[33] Selby R. 'TEG talk': Expanding clinical roles for thromboelastography and rotational thromboelastometry. Hematology (United States). 2020;**20**(1):67-75. DOI: 10.1182/HEMATOLOGY.2020000090

[34] Moore HB et al. Hyperfibrinolysis, physiologic fibrinolysis, and fibrinolysis shutdown: The spectrum of postinjury fibrinolysis and relevance to antifibrinolytic therapy. Journal of Trauma and Acute Care Surgery. 2014;**77**(6):811-817. DOI: 10.1097/TA.0000000000000341

[35] Gomez-Builes JC, Acuna SA, Nascimento B, Madotto F, Rizoli SB. Harmful or physiologic: Diagnosing fibrinolysis shutdown in a trauma cohort with rotational thromboelastometry. Anesthesia and Analgesia. 2018;**127**(4):840-849. DOI: 10.1213/ANE.0000000000003341

[36] Chapman MP et al. The 'death diamond': Rapid thrombelastography identifies lethal hyperfibrinolysis. Journal of Trauma and Acute Care Surgery. 2015;**79**(6):925-929. DOI: 10.1097/TA.0000000000000871

[37] Farrell MS et al. 'Death diamond' tracing on Thromboelastography as a marker of poor survival after trauma. The American Surgeon. 2022;**88**(7):1689-1693. DOI: 10.1177/0003134821998684

[38] Levrat A et al. Evaluation of rotation thrombelastography for the diagnosis of hyperfibrinolysis in trauma patients. British Journal of Anaesthesia. 2008;**100**(6):792-797. DOI: 10.1093/bja/aen083

[39] Stettler GR et al. Redefining postinjury fibrinolysis phenotypes using two viscoelastic assays. Journal of Trauma and Acute Care Surgery. 2019;**86**(4):679-685. DOI: 10.1097/TA.0000000000002165

[40] Kuiper GJAJM et al. Validation of a modified thromboelastometry approach to detect changes in fibrinolytic activity. Thrombosis Journal. 2016;**14**(1):1-13. DOI: 10.1186/s12959-016-0076-2

[41] Kupesiz A et al. Tissue plasminogen activator induced fibrinolysis: Standardization of method using thromboelastography. Blood Coagulation & Fibrinolysis. 2010;**21**(4):320-324. DOI: 10.1097/MBC.0b013e32833464e9

[42] Raza I et al. The incidence and magnitude of fibrinolytic activation in trauma patients. Journal of Thrombosis and Haemostasis. 2013;**11**(2):307-314. DOI: 10.1111/jth.12078

[43] Todd AS. Localization of fibrinolytic activity in tissues. British Medical Bulletin. 1964;**20**(3):210-212. DOI: 10.1093/oxfordjournals.bmb.a070333

[44] Davies AJ, Strachan CJL, Hurlow RA, Stuart J. Fibrinolytic activity of tissue surfaces during surgery. Journal of Clinical Pathology. 1979;**32**(8):822-825. DOI: 10.1136/jcp.32.8.822

[45] Surette AP, Madureira PA, Phipps KD, Miller VA, Svenningsson P, Waisman DM. Regulation of fibrinolysis by S100A10 in vivo. Blood. 2011;**118**(11):3172-3181. DOI: 10.1182/blood-2011-05-353482

[46] Gall LS et al. The S100A10 pathway mediates an occult Hyperfibrinolytic subtype in trauma patients. Annals of Surgery. 2019;**269**(6):1184-1191. DOI: 10.1097/SLA.0000000000002733

[47] Aspelin T, Eriksen M, Lindgaard AK, Lyberg T, Ilebekk A. Cardiac fibrinolytic capacity is markedly increased after brief periods of local myocardial ischemia, but declines following successive periods in anesthetized pigs. Journal of Thrombosis and Haemostasis. 2005;**3**(9):1947-1954. DOI: 10.1111/j.1538-7836.2005.01514.x

[48] Moore HB et al. Fibrinolysis shutdown in trauma: Historical review and clinical implications. Anesthesia and Analgesia. 2019;**129**(3):762-773. DOI: 10.1213/ANE.0000000000004234

[49] Franchini M, Zaffanello M, Mannucci PM. Bleeding disorders in primary fibrinolysis. International Journal of Molecular Sciences. 2021;**22**(13):7077. DOI: 10.3390/ijms22137027

[50] Weisel JW, Litvinov RI. Mechanisms of fibrinolysis and basic principles of management. Management. In: Saba HI, Roberts HR, editors. Hemostasis and Thrombosis. Feb 2014. DOI: 10.1002/9781118833391.ch13

[51] Lijnen HR, Collen D. Mechanisms of physiological fibrinolysis. Baillière's Clinical Haematology. 1995;**8**(2):277-290. DOI: 10.1016/S0950-3536(05)80268-9

[52] Moore HB, Moore BA, Sauaia A, Moore EE. Clinical relevance and practical assessment of fibrinolysis shutdown. ANZ Journal of Surgery. 2020;**90**(4):413-414. DOI: 10.1111/ans.15426

[53] Cardenas JC, Matijevic N, Baer LA, Holcomb JB, Cotton BA, Wade CE. Elevated tissue plasminogen activator and reduced plasminogen activator inhibitor promote hyperfibrinolysis in trauma patients. Shock. 2014;**41**(6):514-521. DOI: 10.1097/SHK.0000000000000161

[54] Chapman MP et al. Overwhelming tPA release, not PAI-1 degradation, is responsible for hyperfibrinolysis in severely injured trauma patients. Journal of Trauma and Acute Care Surgery. 2016;**80**(1):16-25. DOI: 10.1097/TA.0000000000000885

[55] Ozturk M, Sengul N, Dagli M, Kosar A, Bavbek N. Global fibrinolytic capacity in colorectal cancer: A new clue to occult fibrinolysis. Clinical and Applied Thrombosis/Hemostasis. 2003;**9**(2):151-154. DOI: 10.1177/107602960300900210

[56] Moore HB, Moore EE. Temporal changes in fibrinolysis following injury. Seminars in Thrombosis and

Hemostasis. 2020;**46**(2):189-198. DOI: 10.1055/s-0039-1701016

[57] Reed GL, Houng AK. The contribution of activated factor XIII to fibrinolytic resistance in experimental pulmonary embolism. Circulation. 1999;**99**(2):299-304. DOI: 10.1161/01.CIR.99.2.299

[58] Brinjikji W et al. Mechanisms of fibrinolysis resistance and potential targets for thrombolysis in acute ischaemic stroke: Lessons from retrieved stroke emboli. Stroke and Vascular Neurology. 2021;**6**(4):658-667. DOI: 10.1136/svn-2021-001032

[59] Suzuki Y et al. Fibrinolysis-resistant carbonylated fibrin detected in thrombi attached to the vascular wall of abdominal aortic aneurysms. Scientific Reports. 2020;**10**(1):1-9. DOI: 10.1038/s41598-020-77582-1

[60] Moore HB et al. Reperfusion shutdown: Delayed onset of fibrinolysis resistance after resuscitation from Hemorrhagic shock is associated with increased circulating levels of plasminogen activator Inhibitor-1 and Postinjury complications. Blood. 2016;**128**(22):206

[61] Innes D, Sevitt S. Coagulation and fibrinolysis in seriously injured patients. Journal of Clinical Pathology. 1964;**17**(1):1-13

[62] Chakrabarty R, Hocking ED, Fearnley GR. Reaction patterns to three stresses-electroplexy, surgery, and myocardial infarction-of fibrinolysis and plasma fibrinogen. Journal of Clinical Pathology. 1969;**22**:659-662

[63] Kluft C et al. The postoperative fibrinolytic shutdown: A rapidly reverting acute phase pattern for the fast-acting inhibitor of tissue-type plasminogen activator after trauma. Scandinavian Journal of Clinical and Laboratory Investigation. 1985;**45**(7):605-610. DOI: 10.3109/00365518509155267

[64] D'Angelo A, Kluft C, Verheijen JH, Rijken DC, Mozzi E, Mannucci PM. Fibrinolytic shut-down after surgery: Impairment of the balance between tissue-type plasminogen activator and its specific inhibitor. European Journal of Clinical Investigation. 1985;**15**(6):308-312. DOI: 10.1111/j.1365-2362.1985.tb00277.x

[65] Moore HB et al. Fibrinolysis shutdown is associated with a fivefold increase in mortality in trauma patients lacking hypersensitivity to tissue plasminogen activator. Journal of Trauma and Acute Care Surgery. 2017;**83**(6):1014-1022. DOI: 10.1097/TA.0000000000001718

[66] Bach-Gansmo ET, Halvorsen S, Godal HC, Skjønsberg OH. Impaired clot lysis in the presence of human neutrophil elastase. Thrombosis Research. 1995;**80**(2):153-159. DOI: 10.1016/0049-3848(95)00161-J

[67] Barrett CD, Moore HB, Banerjee A, Silliman CC, Moore EE, Yaffe MB. Human neutrophil elastase mediates fibrinolysis shutdown through competitive degradation of plasminogen and generation of angiostatin. Journal of Trauma and Acute Care Surgery. 2017;**83**(6):1053-1061. DOI: 10.1097/TA.0000000000001685

[68] Barbosa da Cruz D et al. DNA-bound elastase of neutrophil extracellular traps degrades plasminogen, reduces plasmin formation, and decreases fibrinolysis: Proof of concept in septic shock plasma. The FASEB Journal. 2019;**33**(12):14270-14280. DOI: 10.1096/fj.201901363RRR

[69] Bach-Gansmo ET, Halvorsen S, Godal HC, Skjønsberg OH. Degradation

of the α-chain of fibrin by human neutrophil elastase reduces the stimulating effect of fibrin on plasminogen activation. Thrombosis Research. 1994;**75**(3):307-317. DOI: 10.1016/0049-3848(94)90241-0

[70] Dow N, Coleman JR, Moore H, Osborn ZT, Sackheim AM, Hennig G, et al. Dense and dangerous: The tissue plasminogen activator-resistant fibrinolysis shutdown phenotype is due to abnormal fibrin polymerization. Journal of Trauma and Acute Care Surgery. 2020;**88**(2):258-265. DOI: 10.1097/TA.0000000000002554. PMID: 31999655; PMCID: PMC7546039

[71] Collet JP et al. Altered fibrin architecture is associated with hypofibrinolysis and premature coronary atherothrombosis. Arteriosclerosis, Thrombosis, and Vascular Biology. 2006;**26**(11):2567-2573. DOI: 10.1161/01. ATV.0000241589.52950.4c

[72] Collet JP et al. The αC domains of fibrinogen affect the structure of the fibrin clot, its physical properties, and its susceptibility to fibrinolysis. Blood. 2005;**106**(12):3824-3830. DOI: 10.1182/blood-2005-05-2150

[73] Mosnier LO, Meijers JCM, Bouma BN. Regulation of fibrinolysis in plasma by TAFI and protein C is dependent on the concentration of thrombomodulin. Thrombosis and Haemostasis. 2001;**85**(1):5-11. DOI: 10.1055/s-0037-1612655

[74] Mosnier LO, Bouma BN. Regulation of fibrinolysis by thrombin Activatable fibrinolysis inhibitor, an unstable carboxypeptidase B that unites the pathways of coagulation and fibrinolysis. Arteriosclerosis, Thrombosis, and Vascular Biology. 2006;**26**(11):2445-2453. DOI: 10.1161/01. ATV.0000244680.14653.9a

[75] Coats TJ, Morsy M. Biological mechanisms and individual variation in fibrinolysis after major trauma. Emergency Medicine Journal. 2020;**37**(3):135-140. DOI: 10.1136/emermed-2019-209181

[76] Macintyre IMC et al. Plasma fibrinolysis and postoperative deep vein thrombosis. The British Journal of Surgery. 1976;**63**:694-697

[77] Roberts DJ et al. Time course and outcomes associated with transient versus persistent fibrinolytic phenotypes after injury: A nested, prospective, multicenter cohort study. Journal of Trauma and Acute Care Surgery. 2019;**86**(2):206-213. DOI: 10.1097/TA.0000000000002099

[78] Leeper CM, Strotmeyer SJ, Neal MD, Gaines BA. Window of opportunity to mitigate trauma-induced coagulopathy: Fibrinolysis shutdown not prevalent until 1 hour post-injury. Annals of Surgery. 2019;**270**(3):528-534. DOI: 10.1097/SLA.0000000000003464

[79] Rossetto A, Vulliamy P, Lee KM, Brohi K, Davenport R. Temporal transitions in fibrinolysis after trauma: Adverse outcome is principally related to late Hypofibrinolysis. Anesthesiology. 2022;**136**(1):148-161. DOI: 10.1097/ALN.0000000000004036

[80] Nicolau-Raducu R et al. Fibrinolysis shutdown is associated with thrombotic and Hemorrhagic complications and poorer outcomes after liver transplantation. Liver Transplantation. 2019;**25**(3):380-387. DOI: 10.1002/lt.25394

[81] Moore HB et al. Detection of early allograft dysfunction at 30 min of reperfusion in liver transplantation: An intraoperative diagnostic tool with real time assessment of graft function. American Journal of Surgery.

2020;**220**(6):1518-1525. DOI: 10.1016/j.amjsurg.2020.08.018

[82] Geerts WH, Code KI, Jay RM, Chen E, Szlai JP. A prospective study of venous thromboembolism after major trauma. The New England Journal of Medicine. 1994;**331**(24):1601-1606

[83] Schultz DJ et al. Incidence of asymptomatic pulmonary embolism in moderately to severely injured trauma patients. The Journal of Trauma: Injury, Infection and Critical Care. 2004;**56**(4):727-733. DOI: 10.1097/01.TA.0000119687.23542.EC

[84] Gando S. Microvascular thrombosis and multiple organ dysfunction syndrome. Critical Care Medicine. 2010;**38**(Suppl. 2):S35-S42. DOI: 10.1097/CCM.0b013e3181c9e31d

[85] Hossain MA et al. Post-traumatic thrombotic microangiopathy: What trauma surgeons need to know? Chinese Journal of Traumatology. 2021;**24**(2):69-74. DOI: 10.1016/j.cjtee.2021.01.004

[86] Hamada SR et al. High level of venous thromboembolism in critically ill trauma patients despite early and well-driven thromboprophylaxis protocol. Annals of Intensive Care. 2017;7:1. DOI: 10.1186/s13613-017-0315-0

[87] Sloan IG, Firkin BG. Impaired fibrinolysis in patients with thrombotic or haemostatic defects. Thrombosis Research. 1989;**55**:559-567

[88] Lisman T. Decreased plasma fibrinolytic potential As a risk for venous and arterial thrombosis. Seminars in Thrombosis and Hemostasis. 2017;**43**(2):178-184. DOI: 10.1055/s-0036-1585081

[89] Raveh Y et al. Fibrinolytic shutdown is associated with intraoperative thrombosis and Hemorrhage during visceral transplant. Seminars in Cardiothoracic and Vascular Anesthesia. 2019;**23**(3):300-308. DOI: 10.1177/1089253219848441

[90] Nelson JT et al. High rate of fibrinolytic shutdown and venous thromboembolism in patients with severe pelvic fracture. The Journal of Surgical Research. 2020;**246**:182-189. DOI: 10.1016/j.jss.2019.09.012

[91] Johna S, O'Callaghan T, Cemaj S, Catalano R. Effect of tissue injury on D-dimer levels: A prospective study in trauma patients. Medical Science Monitor. 2002;**8**(1):5-9

[92] Booth NA. Fibrinolysis and thrombosis, Bailliere's best Pract. Res. Clinics in Haematology. 1999;**12**(3):423-433. DOI: 10.1053/beha.1999.0034

[93] Prior S, Park M, Mann K, Butenas S. Endogenous Procoagulant activity in trauma patients and its relationship to trauma severity. TH Open. 2019;**03**(01):e10-e19. DOI: 10.1055/s-0038-1677030

[94] Chandler WL, Dunbar NM. Thrombin generation in trauma patients. Transfusion. 2009;**49**(12):2652-2660. DOI: 10.1111/j.1537-2995.2009.02335.x

[95] Wolberg AS, Campbell RA. Thrombin generation , fibrin clot formation and hemostasis. Transfusion and Apheresis Science. 2008;**38**:15-23. DOI: 10.1016/j.transci.2007.12.005

[96] Valladolid C et al. Modulating the rate of fibrin formation and clot structure attenuates microvascular thrombosis in systemic inflammation. Blood Advances. 2020;**4**(7):1340-1349. DOI: 10.1182/bloodadvances.2020001500

[97] Wohner N. Role of cellular elements in Thrombus formation and dissolution.

Cardiovascular & Hematological Agents in Medicinal Chemistry. 2008;**6**(3):224-228. DOI: 10.2174/187152508784871972

[98] Wohner N et al. Lytic resistance of fibrin containing red blood cells. Arteriosclerosis, Thrombosis, and Vascular Biology. 2011;**31**(10):2306-2313. DOI: 10.1161/ATVBAHA.111.229088

[99] Wolberg AS. Plasma and cellular contributions to fibrin network formation, structure and stability. Haemophilia. 2010;**16**(Suppl. 3):7-12. DOI: 10.1111/j.1365-2516.2010.02253.x

[100] Fay WP, Eitzman DT, Shapiro AD, Madison EL, Ginsburg D. Platelets inhibit fibrinolysis in vitro by both plasminogen activator inhibitor-1-dependent and -independent mechanisms. Blood. 1994;**83**(2):351-356. DOI: 10.1182/blood.v83.2.351.351

[101] Suzuki Y, Sano H, Mochizuki L, Honkura N, Urano T. Activated platelet-based inhibition of fibrinolysis via thrombin-activatable fibrinolysis inhibitor activation system. Blood Advances. 2020;**4**(21):5501-5511. DOI: 10.1182/bloodadvances.2020002923

[102] Undas A, Ariëns RAS. Fibrin clot structure and function: A role in the pathophysiology of arterial and venous thromboembolic diseases. Arteriosclerosis, Thrombosis, and Vascular Biology. 2011;**31**(12). DOI: 10.1161/ATVBAHA.111.230631

[103] Longstaff C et al. Mechanical stability and fibrinolytic resistance of clots containing fibrin, DNA, and histones. The Journal of Biological Chemistry. 2013;**288**(10):6946-6956. DOI: 10.1074/jbc.M112.404301

[104] Gould TJ et al. Cell-free DNA modulates clot structure and impairs fibrinolysis in Sepsis. Arteriosclerosis, Thrombosis, and Vascular Biology. 2015;**35**(12):2544-2553. DOI: 10.1161/ATVBAHA.115.306035

[105] Goswami J et al. Neutrophil extracellular trap formation and Syndecan-1 shedding are increased after trauma. Shock. 2021;**56**(3):433-439. DOI: 10.1097/SHK.0000000000001741

[106] Zubairova LD et al. Circulating microparticles Alter formation, structure, and properties of fibrin clots. Scientific Reports. 2015;**5**:1-13. DOI: 10.1038/srep17611

[107] Okafor ON, Gorog DA. Endogenous fibrinolysis. Journal of the American College of Cardiology. 2015;**65**(16):1683-1699. DOI: 10.1016/j.jacc.2015.02.040

[108] Meltzer ME et al. Venous thrombosis risk associated with plasma hypofibrinolysis is explained by elevated plasma levels of TAFI and PAI-1. Blood. 2010;**116**(1):113-121. DOI: 10.1182/blood-2010-02-267740

[109] Knudson MM et al. Tissue plasminogen activator resistance is an early predictor of posttraumatic venous thromboembolism: A prospective study from the CLOTT research group. Journal of Trauma and Acute Care Surgery. 2022;**93**(5):597-603. DOI: 10.1097/TA.0000000000003625

[110] Starzl TE, Marchioro TL, Von-Kaula KN, Hermann G, Brittain RS, Waddell WR. Homotransplantation of the liver in humans. Surgery, Gynecology & Obstetrics. 1963;**117**(6):659-676

[111] Groth CG, Pechet L, Starzl TE. Coagulation during and after Orthotopic transplantation of the human liver. Archives of Surgery. 1969;**98**(1):31-34. DOI: 10.1001/archsurg.1969.01340070049006

[112] Dzik WH, Arkin CF, Jenkins RL, Stump DC. Fibrinolysis during liver transplantation in humans: Role of tissue-type plasminogen activator. Blood. 1988;**71**(4):1090-1095. DOI: 10.1182/blood.v71.4.1090.1090

[113] Poon KS et al. Fibrinolysis after reperfusion of liver graft. Acta Anaesthesiologica Taiwanica. 2015;**53**(1):41-43. DOI: 10.1016/j.aat.2014.12.001

[114] De Pietri L, Montalti R, Bolondi G, Serra V, Di Benedetto F. Intraoperative thromboelastography as a tool to predict postoperative thrombosis during liver transplantation. World Journal of Transplantation. 2020;**10**(11):345-355

[115] Xia VW, Steadman RH. Antifibrinolytics in orthotopic liver transplantation: Current status and controversies. Liver Transplantation. 2005;**11**(1):10-18. DOI: 10.1002/lt.20275

[116] Paramo JA, Rifon J, Llorens R, Casares J, Paloma MJ, Rocha E. Intra-and postoperative fibrinolysis in patients undergoing cardiopulmonary bypass surgery. Haemostasis. 1991;**21**:58-64

[117] Mangano DT, Tudor IC, Dietzel C. The risk associated with Aprotinin in cardiac surgery. The New England Journal of Medicine. 2006;**354**(4):353-365. DOI: 10.1056/nejmoa051379

[118] Sussman M et al. Fibrinolysis phenotypes differ amongst cardiac surgery patients: Antifibrinolytic therapy for all? Journal of the American College of Surgeons. 2019;**229**(4):S44-S45. DOI: 10.1016/j.jamcollsurg.2019.08.111

[119] Dhir A. Antifibrinolytics in cardiac surgery. Annals of Cardiac Anaesthesia. 2013;**16**(2):117-125. DOI: 10.4103/0971-9784.109749

[120] Chapman MP et al. Fibrinolysis greater than 3% is the critical value for initiation of antifibrinolytic therapy. Journal of Trauma and Acute Care Surgery. 2013;**75**(6):961-967. DOI: 10.1097/TA.0b013e3182aa9c9f

[121] Benipal S, Santamarina JL, Vo L, Nishijima DK. Mortality and thrombosis in injured adults receiving tranexamic acid in the post-CRASH-2 era. The Western Journal of Emergency Medicine. 2019;**20**(3):443-453. DOI: 10.5811/westjem.2019.4.41698

[122] Karl V, Thorn S, Mathes T, Hess S, Maegele M. Association of Tranexamic Acid Administration with mortality and thromboembolic events in patients with traumatic injury: A systematic review and Meta-analysis. JAMA Network Open. 2022;**5**(3):1-14. DOI: 10.1001/jamanetworkopen.2022.0625

Chapter 5

Cutaneous Microcirculation of the Foot

Gayathri Victoria Balasubramanian and Roozbeh Naemi

Abstract

The skin, the body's largest organ, acts as a shield against infections and injuries. The skin has an inherent ability to autoregulate its blood flow, which depends on extrinsic/intrinsic factors. This function is facilitated by a complex regulatory system that includes local regulation of cutanemicrocirculation involving sensory and autonomic fibres. These play important roles in thermoregulation, maintenance of homeostasis, defence, inflammatory response and nutrition. Any structural or functional damage to the microvasculature can lead to an incongruity in the demand and supply either due to physiological or pathological reasons. Besides, the small fibre nerves supplied by the microvessels can suffer from hypoxia, which in turn can cause problems. By understanding these functional aspects and applying this knowledge for assessment, the complex pathophysiological mechanisms of diseases like Raynaud's and diabetic-foot complications can be better understood. Moreover, microcirculation is crucial for wound healing in both diabetic foot and in pressure ulcers. This chapter aims to discuss the anatomy and physiology of foot microcirculation and its involvement in the pathobiology of certain diseases. Furthermore, various microcirculatory assessment tools and methods are discussed. Acquiring this knowledge can be helpful in providing more effective prevention, diagnosis, and treatment of microcirculatory diseases of the foot.

Keywords: microcirculation, foot, microvessels, skin, perfusion, foot disease

1. Introduction

The microcirculation of the human body plays an important role in the supply of nutrients, protection, elimination of waste products, defence mechanisms and maintaining homeostasis. The skin is the largest and most accessible organ to study microcirculation. The functional aspects of microcirculation are an important factor for foot health for maintaining tissue integrity, responding to noxious and non-noxious stimuli and thermoregulating. There are several pathological conditions in which the foot is adversely affected when the cutaneous microcirculation is compromised. Although microangiopathy, or small vessel disease, is not recognised as a disease in its own right, its involvement in the pathogenesis of diseases such as diabetic foot cannot be denied [1]. An understanding of the structural and functional aspects of microcirculation and the structures, such as small fibre nerves that interact with them, help identify mechanisms through which foot health can be improved. Also, the lack of understanding of

IntechOpen

complications in the presence of complex diseases such as diabetes can be bridged. This, in turn, will help identify diagnostic tools to assess the foot at risk and develop solutions to treat the foot. This chapter aims to provide an overview of the skin of the foot and the neurovascular interactions in the foot. Additionally, it intends to lay a foundation on the microcirculation of the foot, its assessments and applications in understanding the pathological conditions of the foot using selected diseases as an example.

2. The skin of the foot

The skin is the largest organ of the human body measuring about 2 metres in length. The skin can be either glabrous without hair such as in palmar or plantar surfaces or it can be non-glabrous with hair as found in the hand, feet or other regions of the body. Glabrous skin has highly innervated arteriovenous shunts and plays a major role in thermoregulation [2]. In contrast, non-glabrous hairy skin has fewer arteriovenous shunts and is primarily involved in defence and nutrition [2]. Therefore, the skin has both thermoregulatory and non-thermoregulatory roles to maintain homeostasis and preserve human health. Sweating helps to eliminate waste products and toxins. The different types of sweat glands in the skin aid to hydrate and moisturise the skin by the transportation of water; natural moisturising factors such as lactate, urea, sodium and potassium and antimicrobial peptides to the skin surface [3]. The human skin also contains a rich and diverse population of microbial organisms. Many of these microbes inhabit the follicular structures of the skin and their interactions with host cells lead to changes in cell function [4]. At the same time, the skin acts as a physical barrier to prevent injuries and infections due to external stimuli such as chemicals, water, dust, heat, adverse temperature and microorganisms [5]. Lastly, the skin has an intrinsic ability to auto-regulate its blood flow, which depends on some external or internal factors. Such functions are facilitated by a complex regulatory system that includes local regulation of cutaneous microcirculation involving sensory and autonomic fibres [2].

The skin is highly vascular and it is richly supplied by small blood vessels, which are known as microvessels or microvasculature that are a part of the microcirculatory system. The microvasculature is the network of finer arteries, arterioles, capillaries and venules that supply and drain blood from every tissue and organ in the body (**Figure 1**). These microvessels lack a muscular layer and their diameter ranges from 5 to 200 μm [6]. Cutaneous microcirculation plays an important role in the exchange of nutrients in the tissue, in the removal of waste products and last but not least in thermoregulation [7]. Cutaneous blood flow can be substantially altered in response to thermal stress. Vasodilation and increased skin blood flow are essential for heat dissipation during exposure to heat and exercise [8]. Similarly, vasoconstriction and decreased skin blood flow to prevent heat loss to protect against hypothermia during exposure to cold are necessary [8]. These skin blood flow mechanisms both local and reflex are controlled by nerves, endothelial derivatives and metabolic factors [8]. Such responses are a vital aspect of normal thermoregulation. These observations demonstrate that skin blood flow is affected by skin temperature and neurovascular interactions [9]. Hence, it could conceivably be hypothesised that vascular changes due to abnormal neuronal control are reflected in the cutaneous thermal changes. Prior studies have noted that the skin temperature on the feet increases in the presence of complications [10, 11]. This may be key to assessing neurovascular deficits and impairment in microcirculation, which will be discussed later in this chapter.

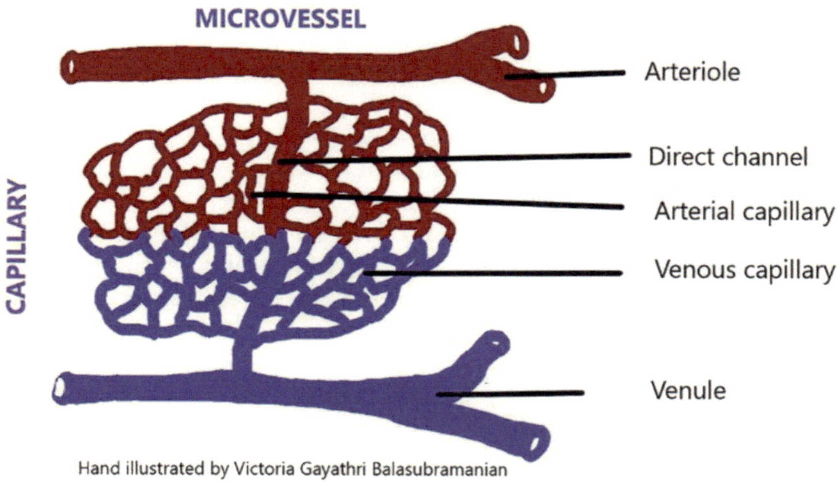

Figure 1.
The microvasculature structure including arterioles, venules and the interconnecting capillaries.

The skin is supplied with both large (>5 μm in diameter) and small fibre nerves (<5 μm in diameter) [12]. The skin's somatosensory system has three basic types of sensory receptors that respond to external stimuli. First, there are mechanoreceptors that respond to mechanical stimuli such as light touch, vibration, pressure and texture. The commonly known mechanoreceptors are Merkel cells or disks, Meissner corpuscles, Ruffini endings and Pacinian corpuscles [13]. These entities are designed to respond to a specific type of mechanical stimuli. For instance, Meissner corpuscles are large, myelinated fibres that detect low-frequency vibration and are present in glabrous (smooth, hairless) skin on fingertips [13]. Pacinian corpuscles contain A-beta fibres, which are rapidly adapting and are present in the deeper layers of skin, ligaments and joints that respond to high-frequency vibration and deep pressure [13]. The Ruffini endings are slow-adapting, encapsulated receptors present in both the glabrous and hairy skin that responds to skin stretch [14, 15]. Finally, Merkel cells, also known as 'touch cells' found in the epidermis of mainly non-glabrous and tactile areas of glabrous skin respond to light touch sensation [16, 17]. Generally, they are low-threshold mechanoreceptors, whilst nociceptors respond to high-threshold stimuli. Second, there are nociceptors found in the skin, joints and viscera that respond to pain induced by a range of factors. Most nociceptors are either C fibres with small diameter unmyelinated axons or A fibres whose axons are myelinated [18–20]. Nociceptors are sensitive to a noxious (harmful) stimulus or a prolonged stimulus that eventually becomes noxious [19–21]. There are different types of nociceptors that respond to chemical, thermal and mechanical stimuli and polymodal that respond to all three [19, 20, 22]. Pain medicated by cutaneous nociceptors can be protective in nature. Stimulation and activation of the terminal branches of the sympathetic and nociceptor fibres result in axon reflex-mediated neurogenic inflammatory reaction, sweating and vasodilation [20, 23]. Third, thermoreceptors respond to thermal stimuli. The Krause end bulbs, Ruffini endings and free nerve endings (the warmth sensation has been attributed to C fibres, whereas cold detection is a function of Aδ fibres) are thermoreceptors [24]. These structures help with thermoregulation and offer protection by sensing harmful thermal stimuli.

As discussed above, the skin is richly supplied by microvessels and nerves. The small fibre nerves supply the microvessels. Therefore, the cutaneous microcirculation is regulated according to the responses of the small fibre nerves to various stimuli. For instance, vasodilation and vasoconstriction to thermoregulate based on temperature are mediated by the small fibres that perceive the signal. Likewise, the small fibre nerves are supplied by the microvessels that are essential for nutrition supply. Any damage to these microvessels results in endoneurial hypoxia, which is attributed to the pathogenesis of diabetic neuropathy [25]. Therefore, the small fibres and the microvasculature are in a mutual relationship in which damage to one or the other has serious consequences, such as loss of a protective sensation or failure of thermoregulation. Thus, in the presence of increased physiological stress from physical stress such as injury or infection and chemical stress, any neurovascular dysfunction impairs microcirculation and loss of tissue integrity [26].

3. Assessment of cutaneous microcirculation of the foot

In recent times, the field of medical research has seen a great deal of innovation in technology to facilitate research, diagnosis, and treatment. This also extends to the area of microcirculation and these microcirculation assessment methods can be categorised into Non-imaging and imaging techniques (**Figure 2**).

Previously, the gold standard method for assessing microcirculation was wire or pressure micromyography of the media thickness to inner lumen ratio (MLR) of subcutaneous small resistance arteries obtained from local biopsies [27]. There is evidence that increased MLR is associated with a poorer prognosis for conditions such as diabetes and hypertension. A reduction in the vascular density in the most distal part of the microcirculation, namely arterioles (100 μm to 7 μm in diameter) and capillaries (around 7 μm in diameter), has been observed in hypertension, diabetes and obesity [27–31]. Conventionally Ankle Brachial Index (ABI) and Toe Brachial Index (TBI) have been used for the diagnosis of peripheral arterial disease (PAD) related to the foot. ABI is an objective diagnostic method but it is considered to be less reliable in cases where the arteries may be calcified as in diabetes, therefore, it is resistant to compression [32, 33]. In such cases, TBI may be used [33–35]. Segmental blood pressure (SEGP) measurements taken at the thigh, calf and ankle are also used to assess PAD, but there are conflicting opinions about its utility [36–38]. However, both ABI and TBI are microcirculatory measurements and do not help to assess the microcirculatory issues. However, recently there are many advanced technologies for non-invasive assessment of microcirculation and neurovascular responses.

3.1 Technologies to assess skin microcirculation

There are currently different devices available to assess microcirculation non-invasively such as laser Doppler flowmetry or fluxmetry (LDF) system, laser Doppler imager (LDI) or laser Doppler perfusion imager (LDPI), laser speckle contrast imager (LSCI) or laser speckle contrast analysis (LASCA), superb microcirculatory imaging (SMI) and transcutaneous oxygen measurement (tcpO2).

3.1.1 Laser Doppler flowmetry or fluxmetry system

An LDF is a device that allows a non-invasive method to monitor changes in the cutaneous peripheral microcirculation [39]. It uses a monochromatic low-energy laser beam penetrating the tissue [39]. The LDF then detects the movement of the red blood cells

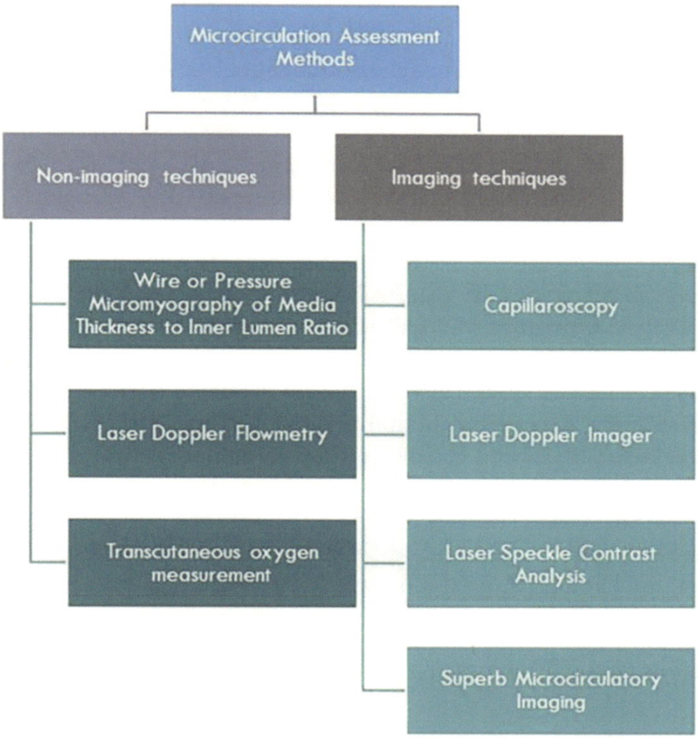

Figure 2.
Microcirculation assessment methods and the two categories of imaging and non-imaging techniques.

(RBCs) in the peripheral microcirculation. Despite the depth to which the laser penetrates being relatively shallow (~1 mm), it is a useful tool for assessing cutaneous microcirculation and its related disorders [40, 41]. However, there are a few disadvantages. Firstly, there may be high spatial variability in the perfusion values because of the regional heterogeneity of skin perfusion and the small measurement area that depends on the contact surface area of single-point LDF probes. This gives a relatively poor reproducibility of the single-point LDF technique [42–44]. Secondly, there may be movement-related artefacts from the attached probes or participants that interfere with the measurements [45, 46]. Thirdly, it is a contact technique and therefore requires the positioning of a probe on the tissues that limit perfusion monitoring of injuries such as an ulcer [42].

3.1.2 Laser Doppler Imager or Laser Doppler Perfusion Imager

The LDI or LDPI uses a contactless imaging technique. LDPI uses a special scanner head and a coherent laser light hits the tissue surface from a distance of about 15 cm [44]. A perfusion image is a collection of LDF points (pixels) arranged on a colour-coded map showing areas of low and high perfusion [42, 44]. The LDPI is useful to monitor perfusion changes for a variety of conditions in dermatology, ulcers, burns, skin grafts, Raynaud's syndrome, and cerebral measurements [44, 47–52]. Nevertheless, there are some limitations to the technique. For instance, LDPI requires long acquisition periods, which prevents the study of rapid changes in perfusion when the surface area to scan is large [42].

3.1.3 Laser speckle contrast imager or laser speckle contrast analysis

LSCI is a technique based on the dynamic changes that occur when a tissue is illuminated with coherent light, forming a speckle pattern at the detector, which is the backscattered light as a result of its interaction with RBCs [41, 53]. It can be used as a tool to capture high-resolution blood flow images, visualise perfusion in various tissues and has a wide range of applications, including both non-real-time investigations and real-time intraoperative perfusion monitoring [53, 54]. A limitation of the LSCI is that it is extremely sensitive to motion, and it can detect the minute movements of RBCs. However, different technical and environmental conditions may interfere with microvascular recordings [42]. When there is external noise, the unwanted motion artefacts are inherent in the measured signal and cannot be easily separated [53]. A simple precautionary step against motion artefacts might be able to create a distraction-free and quiet environment for the patient and overcome the challenge of movement artefacts [42]. Based on the consideration that the speckle contrast of the opaque surface only contains information about unwanted movement and not about perfusion, researchers Mahé et al. [42] added an adhesive opaque patch within the field of view. The contrast measured on the exposed skin contains information about both the blood flow and unwanted movement and when the opaque surface contrast is subtracted, with the addition of a linearity correction, only perfusion remains [53].

3.1.4 Superb Microcirculatory Imaging

More recently, there is an SMI ultrasound technique that expands the range of visible blood flow and allows visualization of low microvascular flow. Compared to traditional Doppler technologies, the advantages of SMI are known for providing high frame rates, resolution and sensitivity with fewer motion artefacts [55, 56]. Unlike traditional colour and power Doppler technologies, which eliminate clutter by suppressing low-velocity components, resulting in data loss and consequently reduced visibility of blood flow in smaller vessels, SMI separates these flow signals from overlying tissue motion artefacts while components with low flow rate and provide detail and definition [55]. Also, it uses clutter motion and uses an adaptive algorithm to identify and remove tissue motion and show a more accurate representation of blood flow. This results in a high-resolution ultrasound image in which minute vessels and low-velocity flows can be demonstrated [55, 56]. The device appears to have applications where visualization of the smallest vessels is required such as when evaluating lesions, cysts, inflammatory diseases, tumours, PAD and foot diseases [55–57].

3.1.5 Transcutaneous oxygen measurement

Apart from the above techniques, transcutaneous oxygen measurement (tcpO2) is a simple, non-invasive method that allows the assessment of the amount of oxygen that diffuses from the capillaries through the epidermis to the electrode and thus provides an estimate of the functional ability of the microvessels to deliver oxygen to the tissue and understand the ischaemic status [58]. Measurement of TcPO2 allows the determination of tissue oxygen tension and can help understand tissue perfusion as it can be performed in people with arterial calcification and no pedal pulse [59, 60]. Therefore, TcPO2 measurement can be useful in evaluating wound healing, especially in people with ulcers in relation to diabetic foot disease [58].

3.2 Tests to assess the skin microcirculation capacity

The different devices and techniques available for monitoring and imaging microcirculation were discussed above. In addition to this, it is important to note that there is a range of provocation tests used to study microcirculation. Some of the commonly used stimuli are pressure, heat stress, cold stress, chemicals (iontophoresis) and postural changes. There is usually a before and after measurement whilst using these stimuli to understand the microcirculatory changes that have occurred. Based on the outcome of interest and the physiological or pathological condition under study, these measurements may help to gain an understanding of the microcirculatory aspects at the region of interest. A select few of these tests are briefly discussed below.

3.2.1 Post-occlusive reactive hyperemia

Post-occlusive reactive hyperemia (PORH) uses occlusion as a stimulus and it is known to be primarily an endothelial-dependent process, however, it involves both endothelial-dependent and independent mechanisms [61, 62]. It helps to assess reperfusion after an occlusion, which is a testing of the tissue's intrinsic ability. The hyperemic response is generated because of shear stress, the tangential frictional force acting at the endothelial cell surface caused by arterial occlusion [63]. The endothelium releases vasodilating substances in response to the mechanical stimulus, which in this case is occlusive pressure [61]. Several factors are known to contribute to vasodilation, which are myogenic, neurogenic, humoral and other local factors such as potassium ions, hydrogen ions, carbon dioxide, catecholamines, prostaglandins and adenosine [61, 62]. It is well known that endothelial nitric oxide and other endothelium-derived agents such as prostaglandins and endothelium-derived hyperpolarising factors play a role in the mechanism of PORH [63]. Apart from these substances, the sensory nerves contribute to the PORH mechanism, which makes this test suitable to assess neurovascular responses [20, 62–67]. The test involves recording a baseline reading, followed by an occlusion using supra systolic pressure (~180 to 220 mmHg) for a certain period and release of pressure. The protocols widely vary depending on various factors such as region of interest, pre-existing complications and outcomes of interest [63]. However, the more recent literature suggests that PORH can be measured reliably using an occlusion time of as little as 30 seconds [63]. PORH is a quick, easy and useful method to assess microcirculation in the arms and feet. PORH is known to be an indicator of cardiovascular risks and foot complications [62, 68, 69]. Therefore, its potential applications could be in the area of risk assessment, which needs further research.

3.2.2 Laser Doppler Imager Flare

Thermal stress tests can be useful to assess microcirculation. The LDI flare test induces nociceptive stimuli-mediated vasodilation and a neurogenic flare through an axon reflex response involving the C-fibres [20]. The LDI flare area, or the size of the area with a hyperemic response, is known to reflect C-fibre function, and the cutaneous perfusion changes immediately below the heating probe reflect the non-neurogenic components involved and can reflect endothelial function in response to heat [20, 70–72]. The test involves heating the local area of the skin to 44°C for 20 minutes or 6 minutes in a stepwise fashion (44°C for 2 minutes, 46°C for 1 minute and finally 47°C for 3 minutes) in a temperature-controlled environment to evoke the flare, followed by scanning of the site using an LDI to measure the area [38, 39, 54]. The 6 minutes protocol is known to produce a significantly larger and more consistent response [72, 73]. The test can be used to assess microcirculatory impairments. Evidence suggests that reduced neurogenic flare along with microcirculatory dysfunction is observed when assessing using the LDI flare test in people with diabetes [70, 72, 74].

3.2.3 Cold stress test

Similar to heat stress, cold stress can also be used to assess microcirculation. The afferent nerves that mediate pain and thermal perception in the skin and the sympathetic efferent vasoconstrictor aspect are involved in the response. The microcirculatory response to a cold stress test might reflect a sympathetic vasoconstrictor and protective vasodilator events [75, 76]. Initial exposure to cold induces cutaneous vasoconstriction, manifested by reduced cutaneous perfusion, but prolonged exposure to cold increases the perfusion, which is a protective hyperemic vasodilator mechanism [8, 76, 77]. The test usually involves a baseline period, followed by exposure to cold (~8°C) for a specified time, followed by a time taken for the temperature to return to room temperature for 20 minutes [75]. The protocols widely vary based on the region of interest, the purpose of the study and the outcomes measured. The cold provocation test is commonly used to study Raynaud's phenomenon, systematic sclerosis, neurovascular changes in diabetes and other conditions [20, 78–80]. The microcirculatory response to cold stress is known to be impaired in people with diabetes [20, 81, 82].

3.2.4 Skin perfusion pressure

The skin perfusion pressure (SPP) is the blood pressure required to restore microcirculatory or capillary flow after controlled occlusion and subsequent flow return. SPP is measured by gradually reducing the inflation cuff pressure and observing isotope washout, the reappearance of pulsatile flow or the movement of RBCs at the measurement site [83, 84]. The minimum external counter pressure exerted by the pressure cuff on the underlying skin is the SPP, above which skin perfusion ceases [83, 84]. It is measured using various techniques but if using LDF the laser Doppler probe is placed beneath a 5.8-cm-wide blood pressure cuff [83]. However, revascularization to restore perfusion is critical for the treatment and prevention of ischemic ulcers in different clinical guidelines. Studies have shown that the test is beneficial for the diagnosis of critical limb ischemia [85–87]. It also helps to assess the limbs for revascularisation procedures to restore perfusion for the treatment and prevention of ischemic ulcers [84]. In addition, SPP is known to serve as an effective guide for amputation-level decisions and for assessing wound healing [83, 84, 88]. However, since there is no consensus in the existing literature, the cut-off value for the diagnosis of various vascular conditions and the potential of SPP as a screening and diagnostic tool in routine practice need to be determined.

4. Microcirculation and select pathological conditions

Assessing the microcirculation of the foot can be helpful in various pathological conditions. Microcirculation plays an important role not only in maintaining homeostasis and tissue integrity but also in tissue injury and response [41]. Acquiring in-depth knowledge of microcirculatory mechanisms can help to explore core issues in various pathological mechanisms. Microcirculatory aspects may be a missing link in understanding complex cutaneous problems and forging solutions. Especially with developing technologies as discussed above, microcirculatory investigations are more accessible. In this chapter, two pathological conditions are discussed briefly in relation to microcirculation.

4.1 Microcirculation in Raynaud's phenomenon

Raynaud's phenomenon is a functional vascular disease presenting with recurrent episodes of limb ischemia in response to cold and emotional stress [89]. Normally, in response to cold temperatures, the body adapts by restricting blood flow to the skin, through a thermoregulatory mechanism to prevent further loss of body heat and maintain core body temperature [90, 91]. However, in Raynaud's phenomenon, a restriction in blood flow, an increase in alpha-2 adrenergic sensitivity in the digital and cutaneous vessel and vasoconstriction of the arteries of the fingers and skin arterioles occur during cold temperatures and emotional stress [89, 90]. Raynaud's phenomenon can be primary or secondary. While the Raynaud's primary phenomenon is idiopathic, the secondary phenomenon is associated with different aetiologies like connective tissue disorders, such as scleroderma, systemic lupus erythematosus, Sjogren syndrome, drugs antimigraine medications, interferon alpha and beta, cyclosporine, and nonselective beta-blockers, and infections, such as parvovirus B19, cytomegalovirus, hepatitis B, and hepatitis C [89, 90]. The investigation of the cutaneous microcirculation can be an important tool for understanding the complex neuro-immunovascular interactions involving both the autonomic and sensory nervous systems in Raynaud's phenomenon [89]. Early diagnosis can help with planning treatment for a better prognosis. There are studies that utilised non-invasive techniques such as LDF and LASCA to understand the underlying pathophysiological mechanisms in Raynaud's phenomenon in order to develop management strategies [89, 92].

4.2 Microcirculation and diabetic foot disease

Diabetes is a chronic disease that leads to various microcirculatory complications in the eyes, kidneys and feet. Diabetic foot syndrome is one of the most common complications of diabetes. Damage to the soft tissue structure is one of the main causes of diabetic foot ulcers, which can be caused by the interaction of several factors such as vascular disease, neuropathy and trauma [41, 93]. Both structural and functional damages to the microvasculature have been linked to diabetes [62, 94]. This can lead to complications in microcirculation, which plays an important role in tissue injury and inflammation [41]. Previous research suggested that certain microcirculatory responses like pressure-induced vasodilation protect the skin and their impairment may contribute to an increased risk of occlusive and ischemic foot injuries [20, 95, 96]. Understanding and studying complex neurovascular responses in the foot can aid in the prevention and early detection of diabetic foot disease and prevent adverse complications such as ulcers and amputations.

5. Conclusion

This chapter provided an overview of the microcirculation of the foot, the structural and functional aspects. Key concepts for understanding microcirculatory functions such as neurovascular interactions were discussed. The chapter also gave a brief overview of different technologies and methods to assess microcirculation, especially in the foot. A selected number of pathological conditions and the relevance of understanding the microcirculation in these conditions were also highlighted. In summary, microvasculature fulfils various roles and forms an important aspect of maintaining skin integrity and homeostasis. It also forms a major part of thermoregulation and tissue injury response mechanisms. Therefore, the growing body of knowledge in the field of microcirculation can offer not only diagnostic but also prognostic solutions for foot health.

Acknowledgements

The authors would like to acknowledge the Staffordshire University Research studentship obtained by the first author, which supported this study.

Conflict of interest

The authors declare that the research was conducted in the absence of any commercial or financial relationships that could be construed as a potential conflict of interest.

Thanks

The authors would like to thank the Centre for Biomechanics and Rehabilitation Technologies, Staffordshire University for the support.

Author details

Gayathri Victoria Balasubramanian[1,2,3] and Roozbeh Naemi[1*]

1 Centre for Biomechanics and Rehabilitation Technologies, Staffordshire University, Stoke on Trent, UK

2 University of Leicester, Leicester, UK

3 Leicester Diabetes Centre, Leicester, UK

*Address all correspondence to: r.naemi@staffs.ac.uk

References

[1] Tesfaye S, Gill G. v. Small vessel disease: A cause of foot ulceration in the neuropathic foot? Practical Diabetes International. 1997;**14**(3):78-79

[2] Stirban A. Microvascular dysfunction in the context of diabetic neuropathy. Current Diabetes Reports. 2014;**14**(11):1-9

[3] Baker LB. Physiology of sweat gland function: The roles of sweating and sweat composition in human health. Temperature: Multidisciplinary Biomedical Journal. 2019;**6**(3):211. Available from: /pmc/articles/PMC6773238/

[4] Gallo RL. Human skin is the largest epithelial surface for interaction with microbes. Journal of Investigative Dermatology. 2017;**137**(6):1213. Available from: /pmc/articles/PMC5814118/

[5] Natsuga K. Epidermal barriers. Cold Spring Harbor Perspective Medicine. 2014;**4**(4). Available from: /pmc/articles/PMC3968788/

[6] Yamaguchi T, Ishikawa T, Imai Y. Integrated Nano-biomechanics. New York, USA: Elsevier; 2018. pp. 1-297

[7] Boulton AJ, Cavanagh PR, Rayman G, Wukich DK. The Foot in Diabetes. Newark: John. In: Boulton AJ, Cavanagh PR, Rayman G, editors. 5th ed. Newark: John Wiley & Sons, Incorporated; 2020. pp. 89-138

[8] Charkoudian N. Skin Blood Flow in Adult Human Thermoregulation: How It Works, When It Does Not, and Why. Mayo Clinic Proceedings. 2003;**78**(5):603-612 Available from: https://www.sciencedirect.com/science/article/abs/pii/S0025619611619307

[9] Fromy B, Abraham P, Bouvet C, Bouhanick B, Fressinaud P, Saumet JL. Early decrease of skin blood flow in response to locally applied pressure in diabetic subjects. Diabetes. 2002;**51**(4):1214-1217 Available from: http://ezproxy.staffs.ac.uk/login?url=http://search.ebscohost.com/login.aspx?direct=true&db=cmedm&AN=11916947&site=ehost-live

[10] Bharara M, Cobb JE, Claremont DJ. Thermography and Thermometry in the Assessment of Diabetic Neuropathic Foot: A Case for Furthering the Role of Thermal Techniques. The International Journal of Lower Extremity Wounds. 2006;**5**(4):250-260 Available from: http://journals.sagepub.com/doi/10.1177/1534734606293481

[11] Gatt A, Falzon O, Cassar K, Ellul C, Camilleri KP, Gauci J, et al. Establishing differences in thermographic patterns between the various complications in diabetic foot disease. International Journal of Endocrinology. 2018;**2018**:1-7 Available from: https://www.hindawi.com/journals/ije/2018/9808295/

[12] Agostini T, de Scisciolo G, Lazzeri D, Almesberger D, Mori A, Cassardo A, et al. Laser-evoked potentials as a novel tool for sensibility assessment following inferior pedicle breast reduction. European Journal of Plastic Surgery. 2013;**36**(7):423-430

[13] Bajwa a KY. Physiology, Vibratory Sense. StatPearls; 2019 Available from: https://www.statpearls.com/ArticleLibrary/viewarticle/31519

[14] Maeda T, Ochi K, Nakakura-Ohshima K, Youn SH, Wakisaka S. The Ruffini ending as the primary mechanoreceptor in the

periodontal ligament: Its morphology, cytochemical features, regeneration, and development. Critical Reviews in Oral Biology and Medicine. 1999;**10**(3):307-327 Available from: https://pubmed.ncbi.nlm.nih.gov/10759411/

[15] Cobo R, García-Piqueras J, Cobo J, Vega JA. The Human Cutaneous Sensory Corpuscles: An Update. Journal of Clinical Medicine. 2021;**10**(2):227 Available from: https://www.mdpi.com/2077-0383/10/2/227/htm

[16] Barbieri JS, Wanat K, Seykora J. Skin: Basic structure and function. Pathobiology of Human Disease: A Dynamic Encyclopedia of Disease Mechanisms. 2014;**2014**:1134-1144

[17] Eduardo CJ, Thomas B, Alexander JL, Steven B. McKee's Pathology of the Skin. 5th ed. Vol. 51. New York: Elsevier; 2019. Available from: https://www.us.elsevierhealth.com/mckees-pathology-of-the-skin-9780702069833.html

[18] Djouhri L, Lawson SN. Aβ-fiber nociceptive primary afferent neurons: A review of incidence and properties in relation to other afferent A-fiber neurons in mammals. In: Brain Research Reviews. 2004. pp. 131-145

[19] Dubin AE. Patapoutian A. Nociceptors: The sensors of the pain pathway. Journal of Clinical Investigation. 2010;**120**:3760-3772

[20] Balasubramanian G, Vas P, Chockalingam N, Naemi R. A synoptic overview of neurovascular interactions in the foot. Frontiers in Endocrinology (Lausanne). 2010;**11**:308 Available from: https://www.frontiersin.org/article/10.3389/fendo.2020.00308/full

[21] Gangadharan V, Kuner R. Pain hypersensitivity mechanisms at a glance. Vol. 6. DMM Disease Models and Mechanisms. In: Company of Biologists. 2013. pp. 889-895

[22] Pinho-Ribeiro FA, Verri WA, Chiu IM. Nociceptor Sensory Neuron–Immune Interactions in Pain and Inflammation. Trends Immunol. Elsevier Ltd; Jan 2017;**38**(1):5-19. DOI: 10.1016/j.it.2016.10.001. Epub 2016 Oct 25. PMID: 27793571; PMCID: PMC5205568

[23] Gibbons CH, Wang N, Freeman R. Capsaicin induces degeneration of cutaneous autonomic nerve fibers. Annals of Neurology. 2010;**68**(6): 888-898

[24] Torebjörk E, Schmelz M. Chapter 38 single-unit recordings of afferent human peripheral nerves by microneurography. In: Dyck PJ, Thomas PK, editors. Peripheral Neuropathy. 4th ed. Pennsylvania, USA: W.B. Elsevier Saunders; 2005. pp. 1003-1014

[25] Körei AE, Istenes I, Papanas N, Kempler P. Small-fiber neuropathy: A diabetic microvascular complication of special clinical, diagnostic, and prognostic importance. Angiology. 2016;**67**(1):49-57 Available from: http://journals.sagepub.com/doi/10.1177/0003319715583595

[26] Levy BI, Schiffrin EL, Mourad JJ, Agostini D, Vicaut E, Safar ME, et al. Impaired tissue perfusion: A pathology common to hypertension, obesity, and diabetes mellitus. Circulation. 2008;**118**(9):968-976 Available from: https://pubmed.ncbi.nlm.nih.gov/18725503/

[27] Rizzoni D, Mengozzi A, Masi S, Rosei CA, de Ciuceis C, Virdis A. New noninvasive methods to evaluate microvascular structure and function. Hypertension. 2022;**79**(5):874-886

Available from: https://www.ahajournals.org/doi/abs/10.1161/HYPERTENSIONAHA.121.17954

[28] Agabiti-Rosei E, Rizzoni D. Microvascular structure as a prognostically relevant endpoint. Journal of Hypertension. 2017;**35**(5):914-921 Available from: https://pubmed.ncbi.nlm.nih.gov/28355167/

[29] Agabiti-Rosei R, Rizzoni D. The effects of hypertension on the structure of human resistance arteries. In: Lip GY, Hall JE, editors. Comprehensive Hypertension. Philadelphia: Elsevier Mosby; 2007. pp. 579-570

[30] Rizzoni D, Rosei EA. Small artery remodeling in diabetes mellitus. Nutrition, Metabolism, and Cardiovascular Diseases. 2009;**19**(8):587-592 Available from: https://pubmed.ncbi.nlm.nih.gov/19481431/

[31] de Ciuceis C, Rossini C, Porteri E, la Boria E, Corbellini C, Mittempergher F, et al. Circulating endothelial progenitor cells, microvascular density and fibrosis in obesity before and after bariatric surgery. Blood Pressure. 2013;**22**(3):165-172. Available from: https://pubmed.ncbi.nlm.nih.gov/23286244/

[32] Suominen V, Rantanen T, Venermo M, Saarinen J, Salenius J. Prevalence and risk factors of PAD among patients with elevated ABI. European Journal of Vascular and Endovascular Surgery. 2008;**35**(6):709-714 Available from: https://pubmed.ncbi.nlm.nih.gov/18313338/

[33] Park SC, Choi CY, Ha YI, Yang HE. Utility of Toe-brachial index for diagnosis of peripheral artery disease. Archieves in Plastic Surgery. 2012;**39**(3):227. Available from: /pmc/articles/PMC3385338/

[34] Hyun S, Forbang NI, Allison MA, Denenberg JO, Criqui MH, Ix JH. Ankle Brachial Index, Toe Brachial Index, and cardiovascular mortality in participants with and without diabetes mellitus. Journal of Vascular Surgery. 2014;**60**(2):390. Available from: /pmc/articles/PMC4112146/

[35] Høyer C, Sandermann J, Petersen LJ. The toe-brachial index in the diagnosis of peripheral arterial disease. Journal of Vascular Surgery. 2013;**58**(1):231-238 Available from: https://www.sciencedirect.com/science/article/pii/S0741521413007106

[36] Gale SS, Scissons RP, Salles-Cunha SX, Dosick SM, Whalen RC, Pigott JP, et al. Lower extremity arterial evaluation: Are segmental arterial blood pressures worthwhile? Journal of Vascular Surgery. 1998;**27**(5):831-839 Available from: http://www.jvascsurg.org/article/S0741521498702629/fulltext

[37] Hirai M, Shionoya S. Segmental blood pressure of the leg and its clinical use. The Japanese Journal of Surgery. 1978;**8**(2):102-110 Available from: https://pubmed.ncbi.nlm.nih.gov/682389/

[38] Gebuhr P, Soelberg M, Henriksen JH. Segmental blood pressure after total hip replacement. Orthopedics. 1992;**15**(4):488-490

[39] Sarnik S, Hofirek I, Sochor O. Laser Doppler fluxmetry. Biomedical Papers of the Medical Faculty of the University Palacky, Olomouc, Czech Republic. 2007;**151**(1):143-146 Available from: https://pubmed.ncbi.nlm.nih.gov/17690759/

[40] Mrowietz C, Franke RP, Pindur G, Sternitzky R, Jung F, Wolf U. Evaluation of Laser-Doppler-Fluxmetry for the diagnosis of microcirculatory disorders. Clinical Hemorheological Microcirculation. 2019;**71**(2):129-135.

Available from: https://pubmed.ncbi.nlm.nih.gov/30584124/

[41] Balasubramanian G, Chockalingam N, Naemi R. The role of cutaneous microcirculatory responses in tissue injury, inflammation and repair at the foot in diabetes. Frontiers in Bioengineering and Biotechnology. 14 Sep 2021;**9**:732-753. DOI: 10.3389/fbioe.2021.732753. Available from: https://www.frontiersin.org/articles/10.3389/fbioe.2021.732753/full

[42] Mahé G, Durand S, Humeau-Heurtier A, Leftheriotis G, Abraham P. Impact of experimental conditions on noncontact laser recordings in microvascular studies. Microcirculation. 2012;**19**(8):669-675 Available from: https://onlinelibrary.wiley.com/doi/full/10.1111/j.1549-8719.2012.00205.x; https://pubmed.ncbi.nlm.nih.gov/22708898/

[43] Tenland T, Salerud EG, Nilsson GE, Oberg PA. Spatial and temporal variations in human skin blood flow. International Journal of Microcirculation, Clinical and Experimental. 1983;**2**(2):81-90

[44] O'Doherty J, McNamara P, Clancy NT, Enfield JG, Leahy MJ. Comparison of instruments for investigation of microcirculatory blood flow and red blood cell concentration. 2009;**14**(3):4025. Available from: https://www.spiedigitallibrary.org/journals/journal-of-biomedical-optics/volume-14/issue-3/034025/Comparison-of-instruments-for-investigation-of-microcirculatory-blood-flow-and/10.1117/1.3149863.full

[45] Välisuo PO. Optical methods for assessing skin flap survival. In: Meglinski I, editor. Biophotonics for Medical Applications. Sawston, Cambridge: Woodhead Publishing; 2015; pp. 331-346

[46] Gupta AK, Gelb AW. Essentials of neuroanesthesia and neurointensive care: A volume in essentials of anesthesia and critical care. Essentials of Neuroanesthesia and Neurointensive Care: A Volume in Essentials of Anesthesia and Critical Care. 2008;**2008**:1-344 Available from: http://www.sciencedirect.com:5070/book/9781416046530/essentials-of-neuroanesthesia-and-neurointensive-care

[47] Raabe A, van de Ville D, Leutenegger M, Szelényi A, Hattingen E, Gerlach R, et al. Laser Doppler imaging for intraoperative human brain mapping. Neuroimage. 2022;**44**(4):1284-1289. Available from: https://pubmed.ncbi.nlm.nih.gov/19049824/

[48] Kanetaka T, Komiyama T, Onozuka A. Laser Doppler skin perfusion pressure in the assessment of Raynaud's phenomenon. TME Journal. 2023:414-416. Available from: https://www.sciencedirect.com/science/article/pii/S1078588403005963

[49] Jeng J, Bridgeman A, Shivnan L, Burns PT. Laser Doppler Imaging Determines need for Excision and Grafting in Advance of Clinical Judgment: A Prospective Blinded Trial. Burns. Elsevier; Nov 2003;**29**(7):665-670. DOI: 10.1016/s0305-4179(03)00078-0. PMID: 14556723. Available from: https://www.sciencedirect.com/science/article/pii/S0305417903000780?casa_token=hGw2dMJWBT8AAAAA:Br747hbuISSvzOnk8CE7kl1RMqs8NUa1j9TEAHPpotgMyWHmlyYR6sIYkw4xfd0OwkD-iJhd

[50] Malanin K, Havu VK, Kolari PJ. Dynamics of cutaneous laser doppler flux with concentration of moving blood cells and blood cell velocity in legs

with venous ulcers and in healthy legs. Angiology. 2004;**55**(1):37-42

[51] Stü M, Springer C, Paech V, Hermes N, Hoffmann M, Altmeyer P. Increased laser Doppler flow in skin tumors corresponds to elevated vessel density and reactive hyperemia. Wiley Online Library. 2006;**12**(1):1-6. Available from: https://onlinelibrary.wiley.com/doi/abs/10.1111/j.0909-725X.2006.00139.x?casa_token=mIam84FT4SAAAAAA:SfKLNIp_Tfc0g9x8Xhw2N7WEjeCgCZJRGipZvpdtk1u1wf6CLYfRHAT9JYTuxKgLYq7eEpiVSah

[52] Ilias MA, Wårdell K, Stü Cker M, Anderson C, Salerud EG. Assessment of pigmented skin lesions in terms of blood perfusion estimates. Wiley Online Library. 2004;**10**(1):43-49 Available from: https://onlinelibrary.wiley.com/doi/abs/10.1111/j.1600-0846.2004.00052.x?casa_token=6Lu8iDg4PYwAAAAA:n89b1sZ0rEAthgxB5seLVXrlBxYXzgDr6r3yyCojuhGuy4nSOXTxmOMJTndIFctylF6EtChOMFrV

[53] Heeman W, Steenbergen W, van Dam GM, Boerma EC, Heeman W. Clinical applications of laser speckle contrast imaging: A review. Journal of Biomedical Optics. 2019;**24**(8):080901 Available from: https://www.spiedigitallibrary.org/journals/journal-of-biomedical-optics/volume-24/issue-8/080901/Clinical-applications-of-laser-speckle-contrast-imaging-a-review/10.1117/1.JBO.24.8.080901.full

[54] Mangraviti A, Volpin F, Cha J, Cunningham SI, Raje K, Brooke MJ, et al. Intraoperative laser speckle contrast imaging for real-time visualization of cerebral blood flow in cerebrovascular surgery: Results from pre-clinical studies. Scientific Reports. 2020;**10**(1):1-13 Available from: https://www.nature.com/articles/s41598-020-64492-5

[55] Toshiba America Medical Systems. Superb Microvascular Imaging. Tustin: Toshiba America Medical Systems; 2017 Available from: http://www.medical.toshiba.com/

[56] Fu Z, Zhang J, Lu Y, Wang S, Mo X, He Y, et al. Clinical applications of superb microvascular imaging in the superficial tissues and organs: A systematic review. Academic Radiology. 2021;**28**(5):694-703 Available from: http://www.academicradiology.org/article/S1076633220301781/fulltext

[57] Suto Y, Sato W, Yamanaka T, Unuma M, Kobayashi Y, Aokawa M, et al. Utility of superb microvascular imaging in the assessment of foot perfusion in patients with critical limb ischemia. Diagnostics. 2022;**12**(11)

[58] Yip WL. Evaluation of the clinimetrics of transcutaneous oxygen measurement and its application in wound care. International Wound Journal. 2015;**12**(6):625. Available from: http://pmc/articles/PMC7950484/

[59] Allen PI, Goldman M. Skin blood flow: A comparison of transcutaneous oximetry and laser Doppler flowmetry. European Journal of Vascular Surgery. 1987;**1**(5):315-318 Available from: http://ezproxy.staffs.ac.uk/login?url=http://search.ebscohost.com/login.aspx?direct=true&db=cmedm&AN=2971577&site=ehost-live

[60] Cooper K, Majdalany BS, Kalva SP, Chandra A, Collins JD, Francois CJ, et al. ACR Appropriateness Criteria ® Lower Extremity Arterial Revascularization—Post-Therapy Imaging. Journal of the American College of Radiology. 2018;**15**(5):S104-S115. DOI: 10.1016/j.jacr.2018.03.011

[61] Wierzbowska J, Wojtkiewicz S, Zbieć A, Wierzbowski R, Liebert A,

Maniewski R. Prolonged postocclusive hyperemia response in patients with normal-tension glaucoma. Medical Science Monitor. 2014;**20**:2607-2616 Available from: http://ezproxy.staffs.ac.uk/login?url=http://search.ebscohost.com/login.aspx?direct=true&db=cmedm&AN=25502623&site=ehost-live

[62] Lanting SM, Barwick AL, Twigg SM, Johnson NA, Baker MK, Chiu SK, et al. Post-occlusive reactive hyperaemia of skin microvasculature and foot complications in type 2 diabetes. Journal of Diabetes and its Complications. 2017;**31**(8):1305-1310. DOI: 10.1016/j.jdiacomp.2017.05.005

[63] Balasubramanian G, Chockalingam N, Naemi R. A systematic evaluation of cutaneous microcirculation in the foot using post-occlusive reactive hyperemia. Microcirculation. 2021;**28**(5):e12692. Available from: http://www.ncbi.nlm.nih.gov/pubmed/33655651

[64] Lorenzo S, Minson CT. Human cutaneous reactive hyperaemia: Role of BKCa channels and sensory nerves. Journal of Physiology. 2007;**585**(Pt 1): 295-303. Available from: http://www.ncbi.nlm.nih.gov/pubmed/17901123

[65] Marche P, Dubois S, Abraham P, Parot-Schinkel E, Gascoin L, Humeau-Heurtier A, et al. Neurovascular microcirculatory vasodilation mediated by C-fibers and Transient receptor potential vanilloid-type-1 channels (TRPV 1) is impaired in type 1 diabetes. Scientific Reports. 2017;**7**(1). Available from: http://www.nature.com/articles/srep44322

[66] Larkin SW, Williams TJ. Evidence for sensory nerve involvement in cutaneous reactive hyperemia in humans. Circulation Research. 1993;**73**(1):147-154 Available from: https://www.ahajournals.org/doi/10.1161/01.RES.73.1.147

[67] Barwick AL, Tessier JW, de Jonge X, Ivers JR, Chuter VH. Peripheral sensory neuropathy is associated with altered postocclusive reactive hyperemia in the diabetic foot. BMJ Open Diabetes Research & Care. 2016;**4**(1):e000235-e000235 Available from: http://drc.bmj.com/lookup/doi/10.1136/bmjdrc-2016-000235

[68] Bittner HB, Diemel KD, Friedel N, Stengele B, Hetzer R. Skin microcirculation and Laser Doppler blood flow in patients supported by the Berlin heart biventricular assist device. Vasa - Journal of Vascular Diseases. 1992;**21**(2):149-153

[69] Yamamoto-Suganuma R, Aso Y. Relationship between post-occlusive forearm skin reactive hyperaemia and vascular disease in patients with Type 2 diabetes-a novel index for detecting micro- and macrovascular dysfunction using laser Doppler flowmetry. Diabetic Medicine. 2009;**26**(1):83-88. Available from: http://doi.wiley.com/10.1111/j.1464-5491.2008.02609.x

[70] Green AQ, Krishnan S, Finucane FM, Rayman G. Altered C-fiber function as an indicator of early peripheral neuropathy in individuals with impaired glucose tolerance. Diabetes Care. 2010;**33**(1):174-176

[71] Vas PRJ, Rayman G. The rate of decline in small fibre function assessed using axon reflex-mediated neurogenic vasodilatation and the importance of age related centile values to improve the detection of clinical neuropathy. PLoS One. 2013;**8**(7):e69920 Available from: https://dx.plos.org/10.1371/journal.pone.0069920

[72] Vas PRJ, Green AQ, Gerry R. Small fibre dysfunction, microvascular

complications and glycaemic control in type 1 diabetes: A case–control study. Diabetologia. 2012;**55**(3):795-800 Available from: http://link.springer.com/10.1007/s00125-011-2417-9

[73] Vas PRJ, Rayman G. Validation of the modified LDIFlare technique: A simple and quick method to assess C-fiber function. Muscle & Nerve. 2013;**47**(3):351-356

[74] Nouri MN, Ahmed A, Bril V, Orszag A, Ng E, Nwe P, et al. Diabetic neuropathy and axon reflex-mediated neurogenic vasodilatation in type 1 diabetes. PLoS One. 2012;**7**(4). Available from: https://pubmed.ncbi.nlm.nih.gov/22529938/

[75] Edward PF, Andrew KL, Smith BP, Holden M, Russell GB, Poehling GG. Measurement of hand microvascular blood flow with isolated cold stress testing and laser Doppler fluxmetry. Journal of Hand Surgery. 1993;**18**(1):143-150. Available from: http://www.ncbi.nlm.nih.gov/pubmed/8423302

[76] van den Brande P, de Coninck A, Lievens P. Skin microcirculation responses to severe local cooling. International Journal of Microcirculation Clinical Experiment. 2019;**17**(2):55-60. Available from: http://www.ncbi.nlm.nih.gov/pubmed/9253681

[77] Cheung SS, Daanen HAM. Dynamic Adaptation of the Peripheral Circulation to Cold Exposure: Microcirculation. John Wiley & Sons, Ltd; Jan 2012;**19**(1):65-77. DOI: 10.1111/j.1549-8719.2011.00126.x. PMID: 21851473. Available from: http://doi.wiley.com/10.1111/j.1549-8719.2011.00126.x

[78] Dinsdale G, Herrick A. Vascular diagnostics for Raynaud phenomenon. Journal of Vascular Diagnostic. 2014;**2**:127. Available from: http://www.dovepress.com/vascular-diagnostics-for-raynaud39s-phenomenon-peer-reviewed-article-JVD

[79] Murray AK, Herrick AL, King TA. Laser Doppler imaging: A developing technique for application in the rheumatic diseases. Rheumatology. 2004;**43**(10):1210-1218. Available from: https://academic.oup.com/rheumatology/article-lookup/doi/10.1093/rheumatology/keh275

[80] Waszczykowska A, Goś R, Waszczykowska E, Dziankowska-Bartkowiak B, Jurowski P. Assessment of skin microcirculation by laser Doppler flowmetry in systemic sclerosis patients. Postepy Dermatological Alergology. 2014;**31**(1):6-11. Available from: http://www.ncbi.nlm.nih.gov/pubmed/24683391

[81] Wilson SB, Jennings PE, Belch JJ. Detection of microvascular impairment in type I diabetics by laser Doppler flowmetry. Clinical Physiology. 1992;**12**(2):195-208. Available from: http://www.ncbi.nlm.nih.gov/pubmed/1533824

[82] Chen YJ, Wang CL, Lo SK. Cutaneous microcirculation of diabetic foot: combined isolated cold stress testing and laser Doppler flowmetry study. Journal of Formos Medical Association. 1997;**96**(8):606-612. Available from: http://www.ncbi.nlm.nih.gov/pubmed/9290270

[83] Yamada T, Ohta T, Ishibashi H, Sugimoto I, Iwata H, Takahashi M, et al. Clinical reliability and utility of skin perfusion pressure measurement in ischemic limbs-Comparison with other noninvasive diagnostic methods. Journal of Vascular Surgery. 2008;**47**(2):318-323

[84] Pan X, Chen G, Wu P, Han C, Ho JK. Skin perfusion pressure as a

predictor of ischemic wound healing potential (Review). Biomedical Reports. 2018;**8**:330-334

[85] Castronuovo JJ, Adera HM, Smiell JM, Price RM. Skin perfusion pressure measurement is valuable in the diagnosis of critical limb ischemia. Journal of Vascular Surgery. 1997;**26**(4):629-637. Available from: http://ezproxy.staffs.ac.uk/login?url=http://search.ebscohost.com/login.aspx?direct=true&db=cmedm&AN=9357464&site=ehost-live

[86] Pitts J. Skin perfusion pressure: A case study demonstrating microcirculatory blood flow. Journal of Diagnostic Medical Sonography. 2014;**30**(4):213-216

[87] Tsai FW, Tulsyan N, Jones DN, Abdel-Al N, Castronuovo JJ, Carter SA. Skin perfusion pressure of the foot is a good substitute for toe pressure in the assessment of limb ischemia. Journal of Vascular Surgery. 2000;**32**(1):32-36 Available from: http://linkinghub.elsevier.com/retrieve/pii/S0741521400644841

[88] Urabe G, Yamamoto K, Onozuka A, Miyata T, Nagawa H. Skin perfusion pressure is a useful tool for evaluating outcome of ischemic foot ulcers with conservative therapy. Annals of Vascular Diseases. 2009;**2**(1):21-216

[89] Mosdósi B, Bölcskei K, Helyes Z. Impairment of microcirculation and vascular responsiveness in adolescents with primary Raynaud phenomenon. Pediatric Rheumatology Online Journal. 23 Mar 2018;**16**(1):20. DOI: 10.1186/s12969-018-0237-x

[90] Musa R, Qurie A. Raynaud Disease. StatPearls. 2022. Available from: https://www.ncbi.nlm.nih.gov/books/NBK499833/

[91] Charkoudian N. Mechanisms and modifiers of reflex induced cutaneous vasodilation and vasoconstriction in humans. Journal of Applied Physiology. 1985;**109**(4):1221-1228. Available from: http://www.ncbi.nlm.nih.gov/pubmed/20448028

[92] Rossa A et al. Alteration of microcirculation is a hallmark of very early systemic sclerosis patients: A laser speckle contrast analysis. Clinical Experimental in Rheumatology. 2013;**31**(Suppl 76):109-114. Available from: https://www.clinexprheumatol.org/abstract.asp?a=6441

[93] Armstrong DG, Boulton AJM, Bus SA. Diabetic Foot Ulcers and Their Recurrence. The New England Journal of Medicine. 2017;**376**(24):2367-2375. Available from: https://www.nejm.org/doi/10.1056/NEJMra1615439

[94] Flynn MD, Tooke JE, Michael D. Microcirculation and the diabetic foot. Vascular Medicine Review. 1990;**1**(2):121-138. Available from: http://journals.sagepub.com/doi/10.1177/1358836X9000100204

[95] Koïtka A, Abraham P, Bouhanick B, Sigaudo-Roussel D, Demiot C, Saumet JL. Impaired pressure-induced vasodilation at the foot in young adults with type 1 diabetes. Diabetes. 2004;**53**(3):721-725. Available from: http://ezproxy.staffs.ac.uk/login?url=http://search.ebscohost.com/login.aspx?direct=true&db=cmedm&AN=14988257&site=ehost-live

[96] Vouillarmet J, Josset-Lamaugarny A, Michon P, Saumet JL, Koitka-Weber A, Henni S, et al. Neurovascular response to pressure in patients with diabetic foot ulcer. Diabetes. 2019;**2019**:db180694. Available from: https://diabetes.diabetesjournals.org/content/early/2018/12/20/db18-0694

Chapter 6

DOACs for the Medical Management of Venous Thromboembolism in Adults

Sharath Kommu and Shalini Arepally

Abstract

Venous thromboembolism (VTE) imposes a significant health care burden. Anticoagulation remains the mainstay of treatment for VTE. For decades, warfarin has been the oral anticoagulant of choice for the medical management of VTE; however, the scope and options for managing VTE have been gradually expanding. The coagulation cascade is a complex sequence of steps, and newer agents that act at different levels on this coagulation cascade have been developed. In the past decade, direct oral anticoagulants (DOACs) have proven to be the up-and-coming alternatives as oral agents in the medical management of VTE and have gradually become the first-line agents. Understanding their mechanism of action, uses, advantages, and disadvantages over other anticoagulants will be discussed in the scope of this chapter.

Keywords: direct oral anticoagulant, venous thromboembolism, direct thrombin inhibitor, factor Xa inhibitor, DOAC, VTE, DTI

1. Introduction

Venous thromboembolism (VTE) refers to a thrombus or blood clot in the veins. It is a preventable and treatable medical condition but is often underdiagnosed. It includes deep venous thrombosis (DVT) and pulmonary embolism (PE) [1]. VTE has an estimated incidence rate of 1–2 per 1000 persons every year and is one of the leading causes of disability-adjusted life years (DALYs) lost and associated with significant healthcare costs [2].

Anticoagulation is a mainstay of treatment for the prevention and management of VTE. Unfractionated heparin (UFH) and vitamin K antagonist (VKA) warfarin have been the treatment of choice for the medical management of VTE for several years. These had several limitations, which had to be accepted, as there were no other options. Some of these limitations could be overcome with the advent of various low molecular heparins (LMWHs) and parenteral indirect thrombin inhibitors like fondaparinux. Despite all the progress in the field of anticoagulation for the management of VTE, there was still an unmet need for safe, efficient anticoagulants, particularly those that can be taken orally. Understanding the mechanism of action of heparin, researchers have realized that factor X and factor II in the coagulation cascade can be potential targets for therapeutic anticoagulation. This led to extensive research and the development of a newer generation of anticoagulants, now called the direct acting oral anticoagulants (DOACs). The discovery of these newer antithrombotic agents that act at different levels in the coagulation cascade has helped to further the management of VTE.

2. The nomenclature

Considering they are new compared to the conventional anticoagulants, they were initially called the NOACs, which stands for novel oral anticoagulants [3]. They were sometimes also called non-vitamin K antagonist oral anticoagulants, which also can be abbreviated as NOAC. Another name given to them was target-specific oral anticoagulants (TSOACs). However, as years passed, international societies have reassessed the name and considering their broad mechanism of action, the name DOAC is given [3]. DOACs include both direct thrombin inhibitors (DTIs) and factor Xa inhibitors—these names are given based on the specific step at which they act on the coagulation cascade (see **Figure 1**).

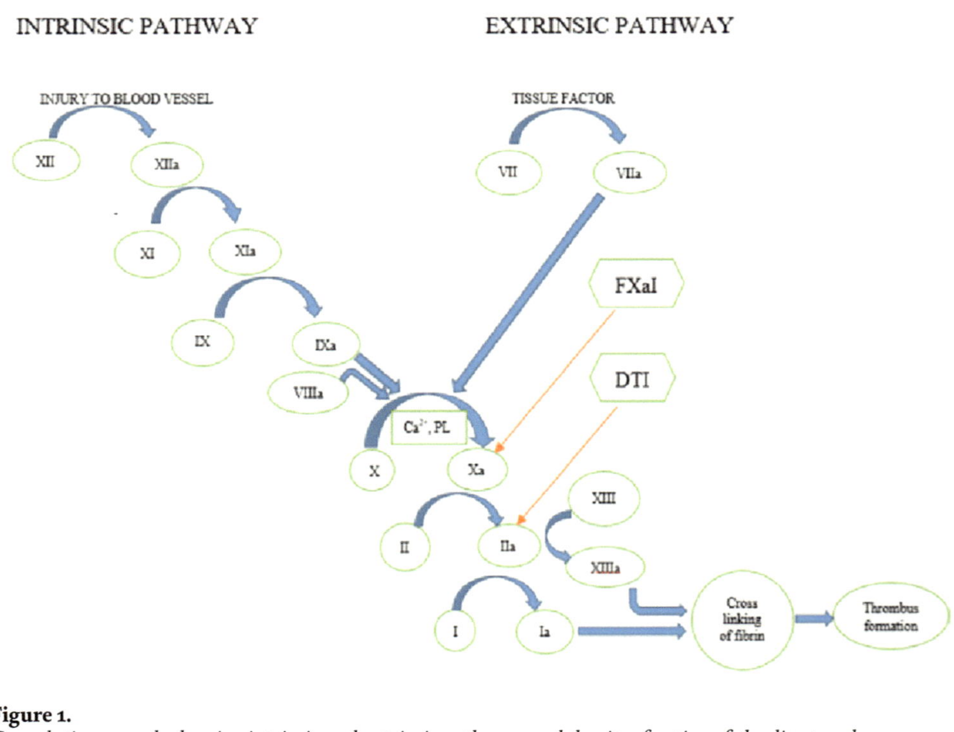

Figure 1.
Coagulation cascade showing intrinsic and extrinsic pathways and the site of action of the direct oral anticoagulants (DOACs). FXaI—factor Xa inhibitor; DTI—direct thrombin inhibitor; PL—platelet membrane phospholipid; Ca^{2+}—calcium ions; II—prothrombin; IIa—thrombin; I—fibrinogen; Ia—fibrin; a—active form of the factor; → inhibiting.

3. Types of DOACs

3.1 Direct thrombin inhibitors

Direct thrombin inhibitors act on activated factor II, also called thrombin (**Figure 1**). Thrombin plays a significant role in the coagulation cascade by helping convert factor I to activated factor I (factor Ia) or fibrin, which further helps form a clot [4]. Investigators working on novel anticoagulants targeted thrombin and designed a new structural class of nonpeptide inhibitors employing 1,2,5-trisubstituted benzimidazole as the central scaffold [5]. Using X-ray structure analysis, they further modified and developed a product named BIBR 1048, which was later named dabigatran, a direct thrombin inhibitor [5]. After initial studies showed the benefit, a larger randomized controlled trial called RE-COVER (Dabigatran vs. Warfarin for the Treatment of Acute Venous Thromboembolism) was conducted and concluded that a fixed dose of dabigatran is as effective as warfarin in treating VTE [6], and later approved for use in VTE. While there are parenteral DTIs like lepirudin, desirudin, bivalirudin, and argatroban, the only oral DTI is dabigatran.

3.2 Activated factor Xa inhibitors

Activated factor X (factor Xa) also plays a critical role in the coagulation cascade as both intrinsic and extrinsic pathways activate it (**Figure 1**). It acts as a catalyst for converting prothrombin to thrombin through the prothrombinase complex, which has factor Xa, activated factor V, Ca^{2+}, and prothrombin on a phospholipid surface. Inhibition of factor Xa can decrease the generation of thrombin, thus decreasing the thrombin-mediated activation of coagulation. Research on factor Xa inhibitors led to the discovery of oxazolidinone derivatives, identified as potent factor Xa inhibitors [7]. This led to the discovery of BAY 59-7939, later named rivaroxaban, a potent and selective direct factor Xa inhibitor with excellent in vivo antithrombotic activity [7]. Other oral factor Xa inhibitors include apixaban, edoxaban, and betrixaban. As one may notice, these generic names end with Xa-ban, highlighting their role as factor Xa inhibitors.

4. Advantages of DOACs over VKA

VKAs like warfarin have been in use for several decades as the anticoagulants of choice; however, they have several limitations. Warfarin requires frequent monitoring to ensure that anticoagulation is within the therapeutic range. This would necessitate the use of additional healthcare resources and carries additional costs. DOACs do not require routine anticoagulation monitoring. Dosing of VKAs is affected by diet, medication interactions, etc., while DOAC dosing is not significantly affected by diet, and they have relatively fewer medication interactions. Warfarin has a long half-life of 35 h, and it takes a few days to reach therapeutic levels, while DOACs have a rapid onset of action. The dosage of warfarin needs to be adjusted regularly based on levels of anticoagulation; on the other hand, DOACs carry a convenient fixed-dose treatment. All these advantages have made DOACs popular and the anticoagulants of choice.

5. Direct oral anticoagulants (DOACs)

DOACs differ in their pharmacokinetic and pharmacodynamic profiles, and it is essential to understand them to select the right drug for the right patient. Here we review the commonly used DOACs and their specific characteristics.

5.1 Oral direct thrombin inhibitors

5.1.1 Dabigatran

Dabigatran was the first DOAC approved by the food and drug administration (FDA). The landmark study that investigated the role of dabigatran on venous thromboembolism is the RE-COVER trial [6]. It was a clinical study that investigated the efficacy and safety of dabigatran for treating acute VTE compared to warfarin. The study was conducted by Boehringer Ingelheim, the manufacturer of dabigatran, and was published in the New England Journal of Medicine in 2009.

The trial enrolled 2564 patients with acute VTE who were randomized to receive either dabigatran or warfarin for 6 months. The primary endpoint was the recurrence of VTE or related death within six months of treatment. It showed that dabigatran was non-inferior to warfarin in preventing recurrent VTE or related death, with 2.4% of patients in the dabigatran group experienced recurrent VTE or related death, compared to 2.1% in the warfarin group [6]. The bleeding rates were similar, with 16.1% of patients in the dabigatran group and 21.9% in the warfarin group [6].

The downsides of this trial include its industry funding, non-inferiority design, the short follow-up period of 6 months, limited generalizability (as this trial excluded those with severe renal impairment, bleeding disorders, etc.), and lack of head-to-head comparison with other novel oral anticoagulants. The follow-up period of 6 months may not be enough and might not capture long-term risks and benefits. Despite these limitations, it was a robust study and showed the benefits of dabigatran.

5.1.1.1 Mechanism of action and pharmacokinetics

It acts as a direct thrombin inhibitor and inhibits free and fibrin-bound thrombin. The prodrug dabigatran etexilate is metabolized to the active drug dabigatran in the liver. It has a bioavailability of 3–7% and a half-life of 12–17 h [8]. It reaches peak plasma levels in 1–2 h and is excreted predominantly in urine [8].

5.1.1.2 Dosage

a. DVT and PE treatment: the dosage of dabigatran in patients with venous thromboembolism is mainly determined by the results of the RE-COVER trial [6]. Based on the results of this trial, FDA approved this drug, and accordingly, the dosage recommended for VTE treatment is:

   - In hemodynamically stable patients—initial five days of therapeutic anticoagulation with a parenteral anticoagulant and then transition to dabigatran at a dosage of 150 mg oral capsule twice daily

b. VTE prophylaxis in total hip arthroplasty (THA) and total knee arthroplasty (TKA): RE-MODEL [9] and RE-NOVATE [10] and RE-NOVATE II [11] trials looked at the effect of dabigatran in knee and hip surgery thromboprophylaxis. A meta-analysis of these trials showed that though both 150 mg and 220 mg can be used for thromboprophylaxis, the higher dose of 220 mg daily is consistently non-inferior to enoxaparin when the surrogate venographic data on major and total venous thromboembolism are evaluated [12]. Hence the following dose is approved and indicated for VTE prophylaxis:

- First dose—110 mg 1–4 h after surgery completion and hemostasis establishment. If not given on the day of surgery, 220 mg can be given after hemostasis is achieved. Maintenance doses 220 mg once daily for 10–14 days.

5.1.1.3 Caution with usage

No large-scale studies compare dabigatran with LMWH in treating malignancy-associated VTE [13]; hence, it is not recommended for treating malignancy-associated VTE. Its use is not recommended in pregnancy and in patients who are breastfeeding. It is not indicated in thrombosis associated with triple-positive antiphospholipid antibody syndrome (APS) [14]. It is mainly eliminated renally (80%); hence, we should be cautious in patients with renal failure. While the recommendation is that no dosage adjustment is required for CrCl > 30, no dosage recommendations are provided, and usage should be avoided in patients with chronic kidney disease with creatinine clearance (CrCl) < or = 30 mL/min or in hemodialysis patients [15]. Premature discontinuation can result in thrombotic events.

5.1.1.4 Adverse reactions

The most common side effect is bleeding, including life-threatening hemorrhage. Other side effects include gastrointestinal like dyspepsia, abdominal discomfort/pain, gastritis, esophagitis, and gastroesophageal reflux disease (GERD). Hypersensitivity reactions like angioedema and anaphylaxis are rare. Agranulocytosis, neutropenia, and thrombocytopenia are also recorded.

5.1.1.5 Specific benefit

While it is better to avoid all DOACs in severe hepatic impairment (transaminases > three times the upper limit of normal or Child-Pugh Class C), dabigatran is preferred to other DOACs in patients with moderate hepatic impairment, however should be used with caution [16].

6. Oral factor Xa inhibitors

6.1 Rivaroxaban

Rivaroxaban is the first factor Xa inhibitor approved for use in VTE. EINSTEIN-VTE study investigators looked at the role of rivaroxaban on VTE [17]. It was a large, randomized, multicenter trial that evaluated the efficacy and safety of oral rivaroxaban compared to standard therapy for treating acute VTE. The study included 4832 patients, randomized to receive either rivaroxaban (15 mg twice daily for three weeks followed by 20 mg once daily) or standard therapy, which consisted of enoxaparin followed by a VKA. The primary efficacy outcome was the composite of symptomatic recurrent VTE or death from any cause during the 3, 6, or 12-month follow-up period. The primary safety outcome was major bleeding. The study found that rivaroxaban was non-inferior to standard therapy for the primary efficacy outcome, with similar rates of recurrent VTE or death from any cause in the rivaroxaban and standard therapy groups (2.1% vs. 3.0%, respectively) [17]. Rivaroxaban was also associated with a significantly lower risk of major bleeding than standard therapy (1.1% vs. 2.2%, respectively) [17]. The study also found that rivaroxaban was associated with a lower risk of clinically relevant non-major bleeding.

The downsides of the Einstein VTE study include lack of blinding as it is an open-labeled study that could lead to reporting bias and a relatively short follow-up period of 12 months. Despite these concerns, the Einstein VTE study is generally regarded as a well-designed and carefully executed trial.

6.1.1 Mechanism of action and pharmacokinetics

It acts by inhibiting factor Xa, thus decreasing the conversion of prothrombin to thrombin. It reaches peak plasma levels in 2–4 h [18]. It is metabolized in the liver and primarily excreted through urine [19]. It has a bioavailability of 66–100% and a half-life of 5–9 h, with a longer half-life in the elderly.

6.1.2 Dosage

a. For VTE treatment: The EINSTEIN VTE study [17] and the study by Buller et al. [20] investigated the dosing of rivaroxaban and as per the results of these studies, the dosage recommended for VTE treatment is:

   - 15 mg twice daily for 21 days, followed by 20 mg once daily.

b. For VTE prophylaxis: Studies by Lassen et al. [21], Turpie et al. [22], and Eriksson et al. [23] studied the dosing of rivaroxaban for hip and knee joint surgery thromboprophylaxis. Based on these studies:

   - In THA and total knee arthroplasty (TKA) patients and in acutely ill hospitalized patients, the dose of rivaroxaban for thromboprophylaxis is 10 mg once daily.

6.1.3 Caution with usage

Rivaroxaban is not recommended with pregnancy or with breastfeeding. As it is primarily excreted in the urine, it is not recommended with CrCl <30 ml/min and should be avoided in CrCl <15 ml/min, in hemodialysis, and peritoneal dialysis patients [19]. Usage should be avoided in moderate to severe hepatic impairment (Child-Pugh Class B or C). It is not recommended for patients with triple-positive APS [14]. Premature discontinuation can result in thrombotic events.

6.1.4 Adverse reactions

Bleeding, including life-threatening hemorrhage, can occur. Other side effects include abdominal pain, cholestasis, increased transaminases, skin rash, Stevens-Johnson syndrome (SJS), hypersensitivity, and anaphylaxis.

6.1.5 Specific benefit

Rivaroxaban may be used in patients with active cancer [16]. It can be used in obesity [24].

6.2 Apixaban

One of the breakthroughs was the AMPLIFY (Apixaban for the Initial Management of Pulmonary Embolism and Deep-Vein Thrombosis as First-Line Therapy) study by Angelli et al., which was a large, randomized, double-blind, non-inferiority trial that compared apixaban to conventional therapy (enoxaparin and warfarin) in the treatment of acute VTE [25]. The study enrolled over 5000 patients from 358 centers in 28 countries and found that apixaban was non-inferior to conventional therapy in preventing recurrent VTE and resulted in significantly less major bleeding. Patients were randomized to receive either apixaban 10 mg twice daily for seven days, followed by 5 mg twice daily for six months, or conventional therapy with enoxaparin followed by warfarin for six months. The results of the study showed that the incidence of recurrent VTE or death related to VTE was significantly lower in the apixaban group compared to the conventional therapy group (2.3% vs. 2.7%), and the incidence of major bleeding was significantly lower in the apixaban group (0.6% vs. 1.8%) [25]. The study concluded that apixaban was non-inferior to conventional therapy for treating acute VTE, with significantly less major bleeding.

Some limitations of this study include the use of a non-inferiority design instead of a superiority design, the fixed-dose regimen used in the study, which may not be optimal for all patients, and the fact that the study was funded by the manufacturer of apixaban, which may raise concerns about potential bias. However, the study followed rigorous scientific standards and was published in the New England Journal of Medicine (NEJM) in 2013.

6.2.1 Mechanism of action and pharmacokinetics

It inhibits factor Xa, thus decreasing the conversion of prothrombin to thrombin. It has an oral bioavailability of approximately 50%. It is metabolized by the liver primarily; excretion is through urine (approximately 27%) and gastrointestinal (GI) tract (biliary and direct intestinal excretion). It has a half-life of approximately 12 h and reaches peak plasma levels in 3–4 h [26].

6.2.2 Dosage

a. For VTE treatment: as per the results of the AMPLIFY study [25], the dosage that is recommended and approved for VTE treatment is:

   - Oral 10 mg twice daily for seven days, followed by 5 mg twice daily.

b. For VTE prophylaxis: the studies by Lassen et al. [27, 28] investigated apixaban for VTE prophylaxis in patients with THA and TKA. Based on these studies, the dosage that is approved and indicated is:

   - In THA and TKA patients: 2.5 mg twice daily for 10–14 days, which can be extended up to 35 days.

6.2.3 Caution with usage

It is not recommended with pregnancy or with breastfeeding. In case of renal impairment, the manufacturer recommends no dosage adjustment. A retrospective cohort study showed that it is safe and effective in patients on dialysis [29]. However, experts advise caution if CrCl < 25%. It should be used cautiously in moderate hepatic impairment (Child-Pugh Class B). Usage must be avoided in severe hepatic impairment (Child-Pugh Class C). It is not recommended for patients with triple-positive APS [14]. As with other DOACs, premature discontinuation can result in thrombotic events.

6.2.4 Adverse reactions

Bleeding, including life-threatening hemorrhage, can occur. Other side effects include nausea, increased transaminases, skin rash, hypersensitivity, and anaphylaxis.

6.2.5 Specific benefit

It may be used in patients with active cancer [30]. It can be used in obesity [24]. Concerning the side effect of bleeding, apixaban carries an advantage over other DOACs because the risk of major or clinically relevant non-major bleeding was less with apixaban than with other DOACs [31].

6.3 Edoxaban

A study by Hokusai—VTE investigators is the landmark trial investigating the role of edoxaban in VTE. It is a randomized, double blind, non-inferiority trial that compared the efficacy and safety of edoxaban versus warfarin in patients with VTE for the initial and extended treatment periods [32]. The study enrolled 8292 patients with acute symptomatic or incidental VTE, and patients were randomized to receive either edoxaban (60 mg once daily or 30 mg once daily in selected patients) or warfarin. The study consisted of two treatment periods: an initial treatment period of 5–10 days and an extended treatment period of up to 12 months. The primary efficacy endpoint was the composite of recurrent symptomatic VTE and VTE-related death during the entire study period. The primary safety endpoint was major bleeding. The study found that edoxaban was non-inferior to warfarin for the primary efficacy endpoint, with a significantly lower rate of major bleeding in the edoxaban group. The 12-month cumulative incidence of the primary efficacy endpoint was 3.2% in the edoxaban group and 3.5% in the warfarin group [32]. The incidence of major bleeding was 1.4% in the edoxaban group and 1.6% in the warfarin group [32]. The study thus showed the beneficial effects of edoxaban for VTE treatment.

The study has limitations as it does not include patients with CrCl less than 30 mL/min. This raises concerns about the generalizability of the results to patients with severe renal impairment or end-stage renal disease, who are at increased risk of bleeding and thromboembolic events. As the manufacturer of edoxaban sponsored the study, it may have introduced bias into the study design and analysis. Despite these shortfalls, it is a significant study that showed the benefits of edoxaban.

6.3.1 Mechanism of action and pharmacokinetics:

It acts by inhibiting factor Xa, thus decreasing the conversion of prothrombin to thrombin. Its bioavailability is approximately 62%. It has a half-life of 10–14 h and reaches peak plasma levels in 1–2 h. It is excreted through urine, with renal clearance being almost 50% of total clearance [33].

6.3.2 Dosage

The dosage recommended and approved for edoxaban is mainly derived from the results of the Honkusai study [32]. After five days of initial therapy with parenteral anticoagulation, we can transition to edoxaban if the patient is hemodynamically stable—60 mg once daily (if patient weight > 60 kg) or 30 mg daily (if patient weight < or = 60 kg). If CrCl 30–50 ml/min—30 mg orally once daily and if the CrCl < 30 ml/min—use is not recommended.

6.3.3 Caution with usage

It is not recommended with pregnancy or with breastfeeding. In renal impairment, it is not recommended if CrCl <30 ml/min [15]. Usage should be avoided in moderate and severe hepatic impairment (Child-Pugh Class B and C). It is not recommended for patients with triple-positive APS [14]. Premature discontinuation can result in ischemic events.

6.3.4 Adverse reactions

Bleeding, including life-threatening hemorrhage. Other side effects include abdominal pain, increased transaminases, skin rash, headache, thrombocytopenia, hypersensitivity, and anaphylaxis.

6.3.5 Specific benefit

Edoxaban can be taken independently of food. It can be given to patients with cancer, except those with GI tract cancers [16].

6.4 Betrixaban

Betrixaban is another factor Xa inhibitor approved by the FDA for VTE prophylaxis in adults hospitalized with acute medical illness [34]. However, it was discontinued in the US in 2020 for business reasons and is not marketed in other countries. Hence, this drug is not extensively discussed here.

7. Proximal vs. distal DVT

The DVTs in the lower extremities can be divided into proximal and distal DVTs. Proximal DVT in the lower extremities is located in the popliteal, femoral, or iliac veins, while distal DVT is usually confined to the calf veins, below the knee, in the anterior tibial, posterior tibial, peroneal and muscular veins.

Whether to treat a distal DVT is controversial, with practices differing from center to center [35]. The consensus is that typically, patients with distal DVT do not need the treatment with anticoagulation and can be followed up with surveillance venous doppler studies. However, factors that favor anticoagulation in cases of distal DVT include the presence of major symptoms, thrombus extension close to the popliteal vein, extensive thrombosis involving multiple veins, unprovoked DVT, presence of persistent risk factors, previous DVT or PE, Corona virus disease 2019 (COVID-19) and patient preference.

8. Duration of treatment for VTE

The decision on the duration of treatment of VTE is complicated and we have to consider various factors. These factors include: the kind of VTE, whether it is provoked or unprovoked, the risk factors—whether they are weak, moderate, or strong risk factors, and whether these risk factors are transient or persistent, and the risk of bleeding associated with anticoagulation. The benefits and risks should be discussed with the patient, and there should be informed decision-making in determining the anticoagulant and treatment duration.

VTEs can be divided into provoked VTEs and unprovoked VTEs. Provoked VTEs occur in the presence of transient or persistent risk factors. The risk factors for VTE can be divided into weak risk factors (odd's ratio (OR) less than two), moderate risk factors (OR 2-9), and strong risk factors (OR more than 10) [36]. Weak risk factors include bed rest over three days, arterial hypertension, type 2 diabetes mellitus, immobility due to sitting, obesity, pregnancy, and varicose veins [36]. Moderate risk factors include arthroscopic knee surgery, autoimmune disease, blood transfusion, central venous lines, intravenous catheters and leads, chemotherapy, congestive heart failure or respiratory failure, hormone replacement therapy, oral contraceptive therapy, postpartum period, infection such as pneumonia, urinary tract infection, inflammatory bowel disease, cancer, stroke, superficial venous thrombosis, and thrombophilia [36]. Strong risk factors include fracture of the lower limb, hospitalization for heart failure or atrial fibrillation, hip or knee replacement, major trauma, myocardial infraction, previous VTE, and spinal cord injury [36].

Usually, if someone has a VTE with a transient risk factor, the duration of treatment is three to six months, and in the presence of VTE which is unprovoked, or in the presence of persistent risk factors, the duration of treatment is longer, possibly lifelong [37].

However, whether to treat those with weak but persistent risk factors with long-term anticoagulation is a matter of debate. There is concern that exposure to long-term anticoagulation may increase the risk of bleeding, particularly in the vulnerable population like the elderly, low body weight, etc. Some experts suggest that in these cases, we can consider reduced dose long-term anticoagulation based on certain studies [38–40].

The AMPLIFY-EXT trial is a study that looked at the effect of reduced-dose apixaban for long-term anticoagulation [38]. It was a randomized, double-blind, placebo-controlled trial that evaluated the efficacy and safety of extended treatment with apixaban in patients who had completed 6–12 months of anticoagulation therapy for VTE. The trial enrolled 2486 patients who were randomly assigned to receive either apixaban 2.5 mg twice daily or a placebo for an additional 12 months of treatment. The trial results showed that extended treatment with apixaban significantly reduced the risk of recurrent VTE or death from any cause, compared to placebo. The incidence of the primary efficacy outcome was 3.8% in the apixaban group and 11.6% in the placebo group, with a hazard ratio of 0.33 (95% confidence interval, 0.22–0.48; P < 0.001) [38]. The incidence of major bleeding was low in both treatment groups, with no significant difference. The incidence of clinically relevant non-major bleeding was higher in the apixaban group than in the placebo group, but this difference was not statistically significant.

The EINSTEIN CHOICE trial looked at the effect of a reduced dose of rivaroxaban for long-term anticoagulation [39]. It was a randomized, double blind, multinational trial that compared the efficacy and safety of rivaroxaban with aspirin for the extended treatment of VTE. The trial enrolled 3365 patients who had completed 6–12 months of anticoagulation therapy for VTE. These patients were randomly assigned to receive one of three regimens for an additional 12 months: rivaroxaban 20 mg once daily, rivaroxaban 10 mg once daily, or aspirin 100 mg once daily. The trial showed that both doses of rivaroxaban were more effective than aspirin for preventing recurrent VTE or VTE-related death. The incidence of the primary efficacy outcome was 1.5% in the rivaroxaban 20 mg group, 1.2% in the rivaroxaban 10 mg group, and 4.4% in the aspirin group, with hazard ratios of 0.34 (95% CI, 0.20–0.59; P < 0.001) and 0.26 (95% CI, 0.14–0.47; P < 0.001), respectively [39]. The incidence of major bleeding was low in all three treatment groups, with no significant differences. The incidence of clinically relevant non-major bleeding was higher in the rivaroxaban group than in the aspirin group.

Based on the results of the above studies, some experts suggest reduced dosing (apixaban 2.5 mg twice daily, rivaroxaban of 10 mg daily) for the maintenance therapy for long-term anticoagulation in patients with weak/mild persistent risk factors and with concerns of bleeding complications.

9. Reversal of the anticoagulation effect of DOACS

There are certain reversal agents that should be considered in case of severe or life-threatening bleeding with DOAC use. These include idarucizumab for dabigatran and andexanet alfa for rivaroxaban, apixaban and edoxaban. It is worth noting that these reversal agents are specific to the respective DOACs, and the effectiveness of reversal may depend on the timing of administration and the severity of the bleeding event. There is some role of prothrombin complex concentrate (PCC) as well, particularly if these first line agents are not available.

9.1 Idarucizumab

It is a reversal agent for dabigatran. It is a monoclonal antibody fragment that binds specifically to dabigatran and neutralizes its anticoagulant effect by forming inactive complexes that are rapidly cleared from the circulation. The RE-VERSE AD study is a phase III, prospective, open-label, multicenter study that evaluated the safety and efficacy of idarucizumab in reversing the anticoagulant effect of dabigatran in patients with uncontrolled bleeding or who required urgent surgery or intervention [41]. The study found that idarucizumab rapidly and effectively reversed the anticoagulant effect of dabigatran in the majority of patients, with a low rate of thrombotic events [41]. The recommended dosage is 5 grams, which is administered as two consecutive 2.5 gm intravenous (IV) bolus injections, given minutes apart [41].

The dose may be repeated if the patient's anticoagulation is not adequately reversed. Though the risk of thrombotic events is low with idarucizumab, we have to carefully monitor the patient for such events.

9.2 Andexanet alfa

Andexanet alfa is a recombinant protein that acts as a decoy receptor for factor Xa inhibitors, such as apixaban, rivaroxaban, and edoxaban. It binds to these anticoagulants in the bloodstream, sequestering them and preventing them from inhibiting the clotting factor Xa. This restores the activity of factor Xa and allows the clotting process to proceed. ANNEXA studies [42, 43] have demonstrated the efficacy and safety of andexanet alfa in reversing the anticoagulant effects of these factor Xa inhibitors in patients experiencing bleeding or requiring urgent surgery or procedures. The recommended dosing for andexanet alfa as per these studies is:

Apixaban: bolus dose of 400 mg over 2 min, followed by a continuous infusion of 4 mg/min for up to 120 min, rivaroxaban: bolus dose of 800 mg over 30 min, followed by a continuous infusion of 8 mg/min for up to 120 min, edoxaban: bolus dose of 800 mg over 30 min, followed by a continuous infusion of 8 mg/min for up to 120 min.

9.3 PCC

Both unactivated prothrombin complex concentrate (uPCC) and activated prothrombin complex concentrate (aPCC) have therapeutic implications and can be used for the reversal of DOACs. The main difference between them is that aPCC contains activated factor VII (FVIIa), while uPCC does not. FVIIa can bypass the normal clotting pathway and activate factor X directly, leading to thrombin generation and clot formation. This makes aPCC a more potent clotting agent compared to uPCC. uPCC is generally preferred for the reversal of factor Xa inhibitors (e.g., apixaban, rivaroxaban, edoxaban) whereas aPCC may be more effective for the reversal of dabigatran [44].

Though there are case reports and case series on the benefits of aPCC in DOAC associated bleeding, there are no high-quality studies to support this. They can be considered if the first line agents are not available. If idarucizumab is not available for reversal of dabigatran, aPCC such as factor eight inhibitor bypassing activity (FEIBA), unactivated 4-factor PCC (contains factors II, VII, IX and X) or 3-factor PCC (contains factors II, IX and X) would be reasonable alternative [44]. If Andexanet alfa is not available for reversal of factor Xa inhibitors, 4-factor PCC is an alternative [44]. If neither is available, 3-factor PCC is the next option.

9.4 Other agents

In case of severe bleeding, in addition to the above reversal agents, antifibrinolytic agents like tranexamic acid, epsilon-aminocaproic acid can be used [44].

10. DOACs for VTE in specific conditions

It is crucial to appropriately dose and be aware of underlying conditions when prescribing DOACs. Some studies suggested that up to 32% of patients experience inappropriate dosing with DOACs [45]. Subtherapeutic and supratherapeutic dosing can expose the patient to the risks of thrombotic or bleeding consequences [46]. Collaboration with pharmacists is critical for evidence-based anticoagulant use and can help avoid inappropriate dosing. Although explained above, under each DOAC, we briefly review some of the special conditions or comorbidities that can influence the selection and dosage of DOACs.

10.1 DOACs for VTE in cancer

Patients with cancer are at an increased risk of VTE. Treating VTE in patients with cancer is complicated due to associated comorbidities like renal failure, being underweight, malabsorption, bleeding risks, etc. Historically LMWHs have been the treatments of choice in cancer VTE. The HOKUSAI -VTE cancer trial demonstrated that edoxaban is non-inferior to dalteparin (LMWH) in preventing recurrent VTE, but there was some increased risk of major bleeding [47]. Select D trial compared rivaroxaban to dalteparin, and there was a decreased risk of VTE and some increased risk of major bleeding [48]. ADAM VTE [49] and Caravaggio trial [50] have demonstrated the benefits of apixaban in cancer VTE patients with comparable major bleeding compared with LMWH. As a deduction from all these studies, evidence and guidelines suggest that edoxaban and rivaroxaban can be used for VTE in cancer patients except those with GI cancers, where there is an increased risk of bleeding with these agents. Apixaban, due to its relatively better major bleeding risk, is preferable to other DOACs in treating VTE in cancer patients [15].

10.2 DOACs for VTE associated with obesity

The large randomized controlled trials investigating different DOACs did not have patient weight as a significant factor in deciding the dosage. In addition, extreme-body weight populations (overweight or underweight) are severely under-represented in these trials [51]. Hence, it was unclear if the same dosages were sufficient for the drug to be effective or to avoid the side effects. The International Society of Thrombosis and Hemostasis analyzed these trials and the sub-group analysis and suggested that DOACs are safe in patients ≤120 kg (BMI ≤ 40 kg/m^2) at standard dosing [15]. For patients with weight > 120 kg or BMI > 40 kg/m^2, it is recommended to avoid dabigatran and edoxaban and can use apixaban and rivaroxaban [24].

10.3 DOACs for VTE associated with low body weight

In patients with low body weight (<60 kg), creatinine clearance is commonly overestimated as the muscle mass is less. In addition, patients with low-body weight can have associated comorbidities like renal impairment, elderly age, etc. Hence, choosing the appropriate DOAC and the correct dosage is crucial to avoid side effects. As there is increased systemic exposure, the dosage of apixaban and edoxaban should be reduced in low-body weight patients (<60 kg), while dabigatran and rivaroxaban should be better avoided [15].

10.4 DOACs for VTE in renal failure

Chronic kidney disease (CKD) patients have an increased risk for thromboembolic and bleeding events. One should be cautious of dosing DOACs appropriately in patients with renal insufficiency. Renal function is one of the most critical determinants of the dosage of DOACs, and renal insufficiency is one of the important factors associated with the inappropriate dosing of DOACs [52]. Among the DOACs, dabigatran is most renally eliminated at 80%, followed by edoxaban (50%), rivaroxaban (35%), and apixaban (27%) [53]. Hence, dabigatran, edoxaban, and rivaroxaban need dosage adjustment with renal impairment and should be avoided in severe renal impairment (CrCl < 30 mL/min). Apixaban undergoes the least renal clearance and is considered the DOAC of choice in severe renal impairment—even in patients with dialysis [29]; however, experts advise caution.

10.5 DOACs for VTE in hepatic failure

DOACs rely on hepatic metabolism for drug clearance, with apixaban accounting for 75%, followed by rivaroxaban (65%), edoxaban (50%), and dabigatran (20%) [54]. Anticoagulation management in patients with advanced chronic liver disease (ACLD) is complicated, particularly as they tend to have bleeding tendencies [55]. They also have an increased risk of thrombosis [55]; a meta-analysis showed a 1.7-fold increase in the risk of thrombosis in patients with cirrhosis compared to the general population [56]. As the large, randomized studies on DOACs have excluded patients with advanced liver failure, we have to rely on small retrospective studies for suggesting the DOACs for patients with advanced liver failure. Liver failure is classified based on the Child-Pugh classification system as Class A, B, and C, and recommendations for DOACs are based on that [55]. While all DOACs can be considered for mild hepatic impairment (Child-Pugh Class A) without dosage adjustment, they should be avoided in severe hepatic impairment (Child-Pugh Class C) [57]. Dabigatran can be considered in patients with moderate hepatic impairment (Child-Pugh Class B), while apixaban, rivaroxaban, and edoxaban should be used with caution [57]. Some guidelines recommend that apixaban and edoxaban should be avoided if transaminases are more than two times the upper limit of normal (ULN), while dabigatran and rivaroxaban should be avoided if transaminases are more than three times ULN [16].

10.6 DOACs for VTE in heart failure

While no specific dosage adjustments are needed for DOACs in heart failure for managing VTE, the presence of heart failure itself is considered a moderate risk factor for venous thromboembolism [36], and heart failure requiring hospitalization is considered a major risk factor for VTE [36]. A study by Fanola et al. showed that the presence of heart failure increased the risk of short-term and long-term VTE, regardless of the presence of other risk factors, and this risk persisted for the 22-year duration of the study in both heart failure with reduced ejection fraction (HFrEF) and heart failure with preserved ejection fraction (HFpEF) [58]. Hence, chronic heart failure would indicate a persistent risk factor, and as discussed earlier in the duration of treatment section, we may have to consider long-term anticoagulation for managing VTE in this population group.

10.7 DOACs for VTE in the elderly

No specific dosage corrections are needed for age in the adult population in managing VTE. The elderly population is at risk of VTE; however, age also is a strong risk factor for anticoagulation-associated bleeding, particularly in those >75 years

[59]. Hence, when treating VTE in the elderly, if long-term anticoagulation is considered (based on unprovoked VTE or persistent risk factors), after the initial 3 to 6 months of treatment, we have to look at the benefits and risks closely. We may have to avoid long-term anticoagulation or consider dose reduction in this population group due to their increased risk of bleeding [59].

10.8 DOACs for VTE in COVID-19

COVID-19 is a prothrombotic, hypercoagulable condition. The prevalence of VTE in COVID-19 ranges from 2.6% to 35.3% [60]. The role of DOACs is not clearly defined in patients with COVID-19. There is evolving evidence on the role of DOACs concerning VTE prophylaxis and treatment in patients with COVID-19. Current guidelines do not recommend DOACs for treating hospitalized patients with acute VTE associated with COVID-19 infection or thromboprophylaxis in acutely ill patients with COVID-19 infection. In both these cases, guidelines recommend using LMWH or UFH [60, 61]. Aslan et al. [62] conducted a study on patients already using DOACs, which did not show any additional benefit of prior DOAC use against intensive care unit (ICU) need or in-hospital mortality. The ACTIV-4B trial by Connors et al. showed that among symptomatic and stable outpatients with COVID-19, thromboprophylaxis with apixaban did not show benefit against all-cause mortality, and venous thrombosis among other parameters studied [63]. In stable patients who are discharged and would benefit from extended prophylaxis, DOACs can be considered [61]. They can also be considered for outpatient management of VTE patients with COVID-19 [61].

11. Future directions

The role of DOACs, which took a boost a decade ago, continues to expand. There are several ongoing trials on the role of DOACs in VTEs. For example—DOAC's role in VTE in patients with ESRD (ClinicalTrials.gov Identifier: NCT04818151), in cancer (ClinicalTrials.gov Identifier: NCT02744092), etc. LEAVE safe with DOACs (ClinicalTrials.gov Identifier: NCT04068727) is a study looking at Clinical Pharmacist intervention to avoid medication errors, which are relatively common with DOACs. Various studies are also in progress looking at the role of DOACs in thromboprophylaxis.

Researchers and scientists continue to explore various other targets in the coagulation cascade to address VTE. Factor XI, a vital component of the intrinsic pathway of the coagulation cascade, is another potential target, as inhibiting factor XI can attenuate thrombosis with little disruption of hemostasis [64]. Recently a study by Verhamme et al. investigated the role of Abelacimab [64]. This monoclonal antibody binds to factor XI and found that it was effective in preventing VTE after TKA [64]. Another agent called Asundexian, a factor XIa inhibitor, is in the investigational stages [65]. Another monoclonal antibody against factor XI, REGN9933, is also in the investigational stages (ClinicalTrials.gov Identifier: NCT05102136).

12. Conclusion

VTE is a relatively common but underdiagnosed healthcare burden. In the past decade, DOACs have revolutionized the management of VTE, overcoming several of the shortcomings of conventional therapies like UFH, LMWH, and VKAs. The oral route of administration, relatively easy dosing, and the lack of need for routine monitoring have made them favorites and preferred anticoagulants of choice in managing VTEs. However, to avoid inadvertent complications like bleeding or worsening of thrombosis and select the appropriate DOAC, dosage, and treatment duration, it is essential to understand their pharmacokinetic properties and be aware of underlying comorbidities and risk factors. With further research on the existing DOACs and prospects of discovering newer anticoagulants, there is hope that the medical management of VTE will be more efficient in the coming times.

Author details

Sharath Kommu* and Shalini Arepally
Marshfield Clinic Health System, Rice Lake, Wisconsin, United States

*Address all correspondence to: sharathkommu@gmail.com

References

[1] Tritschler T, Kraaijpoel N, Gal GL, et al. Venous thromboembolism advances in diagnosis and treatment. Journal of the American Medical Association. 2018;**320**(15):1583-1594. DOI: 10.1001/jama.2018.14346

[2] Scheres LJJ, Lijfering WM, Cannegieter SC. Current and future burden of venous thrombosis: Not simply predictable. Research and Practice in Thrombosis and Haemostasis. 2018;**2**(2):199-208. Published online 2018 Apr 17. DOI: 10.1002/rth2.12101

[3] Barnes GD, Ageno W, Ansell J, et al. Recommendation on the nomenclature for oral anticoagulants: Communication from the SSC of the ISTH. Journal of Thrombosis and Haemostasis. Jun 2015;**13**(6):1154-1156. DOI: 10.1111/jth.12969

[4] Lee CJ, Ansell JE. Direct thrombin inhibitors. British Journal of Clinical Pharmacology. 2011;**72**(4):581-592. DOI: 10.1111/j.1365-2125.2011.03916.x

[5] Hauel NH, Nar H, Priepke H. Structure-based design of novel potent nonpeptide thrombin inhibitors. Journal of Medicinal Chemistry. 2002;**45**(9):1757-1766. DOI: 10.1021/jm0109513

[6] Schulman S, Kearon C, Kakkar AK, et al. Dabigatran versus warfarin in the treatment of acute venous thromboembolism. RE-COVER study group. The New England Journal of Medicine. 2009;**361**:2342-2352. DOI: 10.1056/NEJMoa0906598

[7] Roehrig S, Straub A, Pohlmann J, et al. Discovery of the novel antithrombotic agent 5-chloro-N-({(5S)-2-oxo-3- [4-(3-oxomorpholin-4-yl) phenyl]-1,3-oxazolidin-5-yl}methyl) thiophene2-carboxamide (BAY 59-7939): An oral, direct factor Xa inhibitor. Journal of Medicinal Chemistry. 2005;**48**:5900-5908. DOI: 10.1021/jm050101d

[8] Stangier J. Clinical pharmacokinetics and pharmacodynamics of the oral direct thrombin inhibitor dabigatran etexilate. Clinical Pharmacokinetics. 2008;**2012**(47):285-295. DOI: 10.2165/00003088-200847050-00001

[9] Eriksson BI, Dahl OE, Rosencher N, et al. Oral dabigatran etexilate vs. subcutaneous enoxaparin for the prevention of venous thromboembolism after total knee replacement: The RE-MODEL randomized trial. Journal of Thrombosis and Haemostasis. 2007;**5**(11):2178-2185. DOI: 10.1111/j.1538-7836.2007.02748.x

[10] Eriksson BI, Dahl OE, Rosencher N, et al. Dabigatran etexilate versus enoxaparin for prevention of venous thromboembolism after total hip replacement: A randomised, double-blind, non-inferiority trial. Lancet. 2007;**370**(9591):949-956. DOI: 10.1016/S0140-6736(07)61445-7

[11] Eriksson BI, Dahl OE, Huo MH, et al. Oral dabigatran versus enoxaparin for thromboprophylaxis after primary total hip arthroplasty (RE-NOVATE II). A randomised, double-blind, non-inferiority trial. Thrombosis and Haemostasis. 2011;**105**(4):721-729. DOI: 10.1160/TH10-10-0679

[12] Gómez-Outes A, Terleira-Fernández AI, Suárez-Gea ML, et al. Dabigatran, rivaroxaban, or apixaban versus enoxaparin for thromboprophylaxis after total hip or knee replacement:

Systematic review, meta-analysis, and indirect treatment comparisons. BMJ. 2012;**344**:e3675. DOI: 10.1136/bmj.e3675

[13] Schulman S, Goldhaber SZ, Kearon C, et al. Treatment with dabigatran or warfarin in patients with venous thromboembolism and cancer. Thrombosis and Haemostasis. 2015;**114**(1):150-157. DOI: 10.1160/TH14-11-0977

[14] Pastori D, Menichelli D, Cammisotto V, et al. Use of direct oral anticoagulants in patients with antiphospholipid syndrome: A systematic review and comparison of the international guidelines. Frontiers in Cardiovascular Medicine. 2021;**8**:715878. DOI: 10.3389/fcvm.2021.715878

[15] Chen A, Stecker E, Warden BA. Direct oral anticoagulant use: A practical guide to common clinical challenges. Journal of the American Heart Association. 2020;**9**(13):e017559. DOI: 10.1161/JAHA.120.017559

[16] Streiff MB, Abutalib SA, Farge D, et al. Special series: Approach to the patient with cancer and thrombosis. Update on guidelines for the management of cancer-associated thrombosis. The Oncologist. 2021;**26**(1):e24-e40. DOI: 10.1002/onco.13596

[17] Prins MH, Lensing AW, Bauersachs R, et al. Oral rivaroxaban versus standard therapy for the treatment of symptomatic venous thromboembolism: A pooled analysis of the EINSTEIN-DVT and PE randomized studies. Thrombosis Journal. 2013;**11**(1):21. DOI: 10.1186/1477-9560-11-21

[18] Mueck W, Stampfuss J, Kubitza D, et al. Clinical pharmacokinetic and pharmacodynamic profile of rivaroxaban. Clinical Pharmacokinetics. 2014;**53**(1):1-1. DOI: 10.1007/s40262-013-0100-7

[19] Kumar S, Lim E, Covic A, et al. Anticoagulation in concomitant chronic kidney disease and atrial fibrillation: JACC review topic of the week. Journal of the American College of Cardiology. 2019;**74**:2204-2215. DOI: 10.1016/j.jacc.2019.08.1031

[20] Buller HR, Lensing AWA, Prins MH, et al. A dose-ranging study evaluating once-daily oral administration of the factor Xa inhibitor rivaroxaban in the treatment of patients with acute symptomatic deep vein thrombosis: The Einstein–DVT dose-ranging study. Clinical Trials & Observations. Blood. 2008;**112**(6):2242-2247. DOI: 10.1182/blood-2008-05-160143

[21] Lassen MR, Ageno W, Borris LC, et al. Rivaroxaban versus enoxaparin for thromboprophylaxis after total knee arthroplasty. The New England Journal of Medicine. 2008;**358**(26):2776-2786. DOI: 10.1056/NEJMoa076016

[22] Turpie AG, Lassen MR, Davidson BL, et al. Rivaroxaban versus enoxaparin for thromboprophylaxis after total knee arthroplasty (RECORD4): A randomized trial. Lancet. 2009;**373**(9676):1673-1680. DOI: 10.1016/S0140-6736(09)60734-0

[23] Eriksson BI, Borris LC, Friedman RJ, et al. Rivaroxaban versus enoxaparin for thromboprophylaxis after hip arthroplasty. The New England Journal of Medicine. 2008;**358**(26):2765-2775. DOI: 10.1056/NEJMoa0800374

[24] Sebaaly J, Kelley D. Direct oral anticoagulants in obesity: An updated literature review. The Annals of Pharmacotherapy.

2020;**54**(11):1144-1158. DOI: 10.1177/1060028020923584

[25] Agnelli G, Buller HR, Cohen A, et al. Oral apixaban for the treatment of acute venous thromboembolism. New England Journal of Medicine. 2013;**369**:799-808. DOI: 10.1056/NEJMoa1302507

[26] Byon W, Garonzik S, Boyd RA, et al. Apixaban: A clinical pharmacokinetic and pharmacodynamic review. Clinical Pharmacokinetics. 2019;**58**(10):1265-1279. DOI: 10.1007/s40262-019-00775-z

[27] Lassen MR, Raskob GE, Gallus A, et al. Apixaban or enoxaparin for thromboprophylaxis after knee replacement. The New England Journal of Medicine. 2009;**361**(6):594-604. DOI: 10.1056/NEJMoa0810773

[28] Lassen MR, Gallus A, Raskob GE, et al. Apixaban versus enoxaparin for thromboprophylaxis after hip replacement. The New England Journal of Medicine. 2010;**363**(26):2487-2498. DOI: 10.1056/NEJMoa1006885

[29] Wetmore JB, Herzog CA, Yan H, et al. Apixaban versus warfarin for treatment of venous thromboembolism in patients receiving long-term dialysis. Clinical Journal of the American Society of Nephrology. 2022;**17**(5):693-702. DOI: 10.2215/CJN.14021021

[30] Agnelli G, Becattini C, Meyer G, et al. Apixaban for the treatment of venous thromboembolism associated with cancer. The New England Journal of Medicine. 2020;**382**(17):1599-1607. DOI: 10.1056/NEJMoa1915103

[31] Cohen AT, Hamilton M, Mitchell SA, et al. Comparison of the novel oral anticoagulants apixaban, dabigatran, edoxaban, and rivaroxaban in the initial and long-term treatment and prevention of venous thromboembolism: Systematic review and network meta-analysis. PLoS One. 2015;**10**(12):e0144856. DOI: 10.1371/journal.pone.0144856. eCollection 2015

[32] The Hokusai-VTE Investigators. Edoxaban versus warfarin for the treatment of symptomatic venous thromboembolism. The New England Journal of Medicine. 2013;**369**:1406-1415. DOI: 10.1056/NEJMoa1306638

[33] Parasrampuria DA, Truitt KE. Pharmacokinetics and pharmacodynamics of edoxaban, a non-vitamin K antagonist oral anticoagulant that inhibits clotting factor Xa. Clinical Pharmacokinetics. 2016;**55**:641-655. DOI: 10.1007/s40262-015-0342-7

[34] FDA Approves Betrixaban for Venous Thromboembolism. Available from: https://ashpublications.org/ashclinicalnews/news/3299/FDA-Approves-Betrixaban-for-Venous-Thromboembolism

[35] Stevens SM, Woller SC, Kreuziger LB, et al. Antithrombotic therapy for VTE disease: Second update of the CHEST guideline and expert panel report. Chest. 2021;**160**(6):e545-e608. DOI: 10.1016/j.chest.2021.07.055

[36] Konstantinides SV, Meyer G, Becattini C, et al. 2019 ESC guidelines for the diagnosis and management of acute pulmonary embolism developed in collaboration with the European Respiratory Society (ERS). European Heart Journal. 2020;**41**(4):543-603. DOI: 10.1093/eurheartj/ehz405

[37] Ortel TL, Neumann I, Ageno W, et al. American Society of Hematology 2020 guidelines for management of venous thromboembolism: Treatment of deep vein thrombosis and

pulmonary embolism. Blood Advances. 2020;**4**(19):4693-4738. DOI: 10.1182/bloodadvances.2020001830

[38] Agnelli G, Buller HR, Cohen A, et al. Apixaban for extended treatment of venous thromboembolism. The New England Journal of Medicine. 2013;**368**:699-708. DOI: 10.1056/NEJMoa1207541

[39] Weitz JI, Lensing AWA, Prins MH, et al. Rivaroxaban or aspirin for extended treatment of venous thromboembolism. The New England Journal of Medicine. 2017;**376**:1211-1222. DOI: 10.1056/NEJMoa1700518

[40] Vasanthamohan L, Boonyawat K, Chai-Adisaksopha C, et al. Reduced-dose direct oral anticoagulants in the extended treatment of venous thromboembolism: A systematic review and meta-analysis. Journal of Thrombosis and Haemostasis. 2018;**16**(7):1288-1295. DOI: 10.1111/jth.14156. Epub 2018 Jun 17

[41] Pollack CV, Reilly PA, Eikelboom J, et al. Idarucizumab for dabigatran reversal. The New England Journal of Medicine. 2015;**373**:511-520. DOI: 10.1056/NEJMoa1502000

[42] Siegal DM, Curnutte JT, Connolly SJ, et al. Andexanet alfa for the reversal of factor Xa inhibitor activity. The New England Journal of Medicine. 2015;**373**:2413-2424. DOI: 10.1056/NEJMoa1510991

[43] Connolly SJ, Crowther M, Eikelboom JW, et al. Full study report of andexanet alfa for bleeding associated with factor Xa inhibitors. The New England Journal of Medicine. 2019;**380**:1326-1335. DOI: 10.1056/NEJMoa1814051

[44] Tomaselli GF, Mahaffey KW, Cuker A, et al. 2020 ACC expert consensus decision pathway on management of bleeding in patients on oral anticoagulants. Journal of the American College of Cardiology. 2020;**76**(5):594-622. DOI: 10.1016/j.jacc.2020.04.053

[45] Ruiz Ortiz M, Muniz J, Rana Miguez P, et al. Inappropriate doses of direct oral anticoagulants in real-world clinical practice: Prevalence and associated factors. A subanalysis of the FANTASIIA registry. Europace. 2018;**20**(10):1577-1583. DOI: 10.1093/europace/eux316

[46] Yao X, Shah ND, Sangaralingham LR, et al. Non-vitamin K antagonist oral anticoagulant dosing in patients with atrial fibrillation and renal dysfunction. Journal of the American College of Cardiology. 2017;**69**:2779-2790. DOI: 10.1016/j.jacc.2017.03.600

[47] Raskob GE, Nv E, Verhamme P, et al. Edoxaban for the treatment of cancer-associated venous thromboembolism. The New England Journal of Medicine. 2018;**378**:615-624. DOI: 10.1056/NEJMoa1711948

[48] Young AM, Marshall A, Thirlwall J, et al. Comparison of an oral factor Xa inhibitor with low molecular weight heparin in patients with cancer with venous thromboembolism: Results of a randomized trial (SELECT-D). Journal of Clinical Oncology. 2018;**36**(20):2017-2023. DOI: 10.1200/JCO.2018.78.8034

[49] McBane RD, Wysokinski WE, Le-Rademacher JG, et al. Apixaban and dalteparin in active malignancy-associated venous thromboembolism: The ADAM VTE trial. Journal of Thrombosis and Haemostasis. 2020;**18**(2):411-421. DOI: 10.1111/jth.14662. Epub 2019 Nov 28

[50] Ageno W, Vedovati MC, Cohen A. Bleeding with apixaban and dalteparin

in patients with cancer-associated venous thromboembolism: Results from the Caravaggio study. Thrombosis and Haemostasis. 2021;**121**(5):616-624. DOI: 10.1055/s-0040-1720975

[51] Boonyawat K, Caron F, Li A, et al. Association of body weight with efficacy and safety outcomes in phase III randomized controlled trials of direct oral anticoagulants: A systematic review and meta-analysis. Journal of Thrombosis and Haemostasis. 2017;**15**:1322-1333. DOI: 10.1111/jth.13701

[52] Sanghai S, Wong C, Wang Z, et al. Rates of potentially inappropriate dosing of direct-acting oral anticoagulants and associations with geriatric conditions among older patients with atrial fibrillation: The SAGE-AF study. Journal of the American Heart Association. 2020;**9**:e014108. DOI: 10.1161/JAHA.119.014108

[53] Steffel J, Verhamme P, Potpara TS, et al. The 2018 European heart rhythm association practical guide on the use of non-vitamin K antagonist oral anticoagulants in patients with atrial fibrillation. European Heart Journal. 2018;**39**:1330-1393. DOI: 10.1093/eurheartj/ehy136

[54] Qamar A, Vaduganathan M, Greenberger NJ, et al. Oral anticoagulation in patients with liver disease. Journal of the American College of Cardiology. 2018;**71**:2162-2175. DOI: 10.1016/j.jacc.2018.03.023

[55] Semmler G, Pomej K, Bauer DJM, et al. Safety of direct oral anticoagulants in patients with advanced liver disease. Liver International. 2021;**41**(9):2159-2170. DOI: 10.1111/liv.14992

[56] Ambrosino P, Tarantino L, Di Minno G, et al. The risk of venous thromboembolism in patients with cirrhosis. A systematic review and meta-analysis. Thrombosis and Haemostasis. 2017;**117**(1):139-148. DOI: 10.1160/TH16-06-0450

[57] Ballestri S, Capitelli M, Fontana MC, et al. Direct oral anticoagulants in patients with liver disease in the era of non-alcoholic fatty liver disease global epidemic: A narrative review. Advances in Therapy. 2020;**37**:1910-1932. DOI: 10.1007/s12325-020-01307-z

[58] Fanola CL, Norby FL, Shah AM, et al. Incident heart failure and long-term risk for venous thromboembolism. Journal of the American College of Cardiology. 2020;**75**(2):148-158. DOI: 10.1016/j.jacc.2019.10.058

[59] Gross PL, Chan NC. Thromboembolism in older adults. Frontiers in Medicine. 2021;**7**:470016. DOI: 10.3389/fmed.2020.470016

[60] Schutgens RE. DOAC in COVID-19: Yes or no? Hema. 2021;**5**(1):e526. DOI: 10.1097/HS9.0000000000000526

[61] Flaczyk A, Rosovsky RP, Reed CT, et al. Comparison of published guidelines for management of coagulopathy and thrombosis in critically ill patients with COVID 19: Implications for clinical practice and future investigations. Critical Care. 2020;**24**:559. DOI: 10.1186/s13054-020-03273-y

[62] Aslan B, Akyüz A, Işık F, et al. The effect of chronic DOAC treatment on clinical outcomes of hospitalized patients with COVID-19. International Journal of Clinical Practice. 2021;**75**(9):e14467. DOI: 10.1111/ijcp.14467

[63] Connors JM, Brooks MM, Sciurba FC, et al. Effect of antithrombotic therapy on clinical outcomes in outpatients with clinically

stable symptomatic COVID-19. The ACTIV-4B randomized clinical trial. Journal of the American Medical Association. 2021;**326**(17):1703-1712. DOI: 10.1001/jama.2021.17272

[64] Verhamme P, Yi AB, Segers A, et al. Abelacimab for prevention of venous thromboembolism august 12, 2021. The New England Journal of Medicine. 2021;**385**:609-617. DOI: 10.1056/NEJMoa2105872

[65] First patients enrolled in Bayer's Phase III trial for FXIa inhibitor asundexian - Clinical Trials Arena. Comment. February 10, 2023. Available from: www.clinicaltrialsarena.com/comment/bayer-trial-asundexian/